Practical Natural Gas Engineering

Practical Natural Gas Engineering

R. V. Smith

PennWell Books
PennWell Publishing Company
Tulsa, Oklahoma

Dedicated to
*Maxine, Carol,
and Janet*

PennWell Publishing Company
1421 South Sheridan Road/P. O. Box 1260
Tulsa, Oklahoma 74101

Library of Congress Cataloging in Publication Data

Smith, R. V. (Robert Vincent), 1913–
Practical natural gas engineering.

Includes bibliographical references.
1. Gas, Natural. 2. Gas wells. I. Title.
TN880.S63 1983 622'.3382 82-24684
ISBN 0-87814-225-8

Printed in the United States of America

1 2 3 4 5 87 86 85 84 83

CONTENTS

ACKNOWLEDGMENTS

Special acknowledgment is due Mr. Elmer E. Templeton for his suggestion that a book such as this was needed and for his encouragement and prodding to complete the book. Mr. M.J. Fetkovich reviewed chapters 6, 7, and 8 and made many helpful suggestions, including the suggestion that type curves be used to analyze pressure drawdown tests. Many others made contributions. Also, the privilege of working many years as a natural gas engineer for Phillips Petroleum Company contributed to the writing of the book.

The following organizations graciously gave permission to use material from their publications.

American Gas Association
American Petroleum Institute
Corporation Commission of the State of Oklahoma
Gas Processors Suppliers Association
Interstate Oil Compact Commission
Pacific Energy Association, formerly
California Natural Gasoline Association
Society of Petroleum Engineers of AIME

Practical Natural Gas Engineering

1 Introduction to Natural Gas Engineering

Natural gas engineering is the art of recovering a maximum amount of gas from a reservoir and bringing the gas to the surface at as high a pressure as practical. Both objectives should be accomplished as efficiently and economically as possible. Natural gas engineering also includes forecasting the future performance of reservoirs and the future production from gas wells. The ability to forecast the volumes and the pressures at which the gas is available from wells and reservoirs with an acceptable degree of accuracy is extremely important to the natural gas industry.

Natural gas engineering differs very little from conventional reservoir engineering for oil reservoirs. The principal differences are the units used to describe the production process and the effect of pressure-volume-temperature relationships on the behavior of gas as it flows in porous media and in pipelines. Also, natural gas engineering for the most part is concerned with the single-phase flow of an extremely compressible fluid as compared to oil reservoirs. Nevertheless, it is easy to convert calculations for an all-liquid system to a gaseous system. Progress in natural gas engineering has been hindered to some extent by its preoccupation with the so-called back pressure test and the unavoidable conclusion that gas wells showed evidence of non-Darcy flow.

Although papers published in the 1940s have concluded that gas flow in porous media and gas flow into wells could exhibit non-Darcy behavior, many engineers have taken the position that non-Darcy behavior is a result of the changes in viscosity and compressibility of the gas as it moves to the well. This divergence in views, along with the preoccupation of natural gas engineering with well testing, has proved to be a major stumbling block to natural gas engineering progress. The work of Al-Hussainy, Ramey and Crawford and the work of subsequent investigators have shown that adequately handling changes in viscosity and compressibility still leaves evidence of non-Darcy behavior in the flow of gas to wells.* Further, an understanding of the flow of liquids in porous media has proved that there are no shortcuts to stabilization or magic tests that give the stabilized characteristics of a gas well in a short period of time. Thus, the author believes there is no fundamental difference between the behavior of wells producing liquids and wells producing gas.

The author has attempted to bridge the gap between the results of empirical testing and theory of unsteady-state flow in porous media and thus perhaps to close the gap between conventional reservoir engineering and an understanding of the behavior of gas wells. The latter becomes very important with the recent prices of natural gas and the subsequent push to produce lower-capacity wells at lower-than-ever pressures. The net result has been to increase the importance of natural gas engineering.

Since the author assumes that the reader has some knowledge of reservoir engineering as applied to oil reservoirs and may have had several courses in reservoir engineering, many of the principal subjects in reservoir engineering have not been emphasized. For example, geology has not been mentioned, and the more-technical features of core analysis and results have not been discussed in detail. Instead, the emphasis has been on the behavior of gas wells and gas reservoirs.

RESERVES

The reader will note that the author has avoided the use of the terms *reserves* and *recoverable reserves*. Historically, common practice has been to estimate recoverable gas reserves to be 85% of the gas in place in a dry gas reservoir. Assuming that the West Panhandle and Hugoton fields of Texas had a rock pressure of 435 psig at discovery (shutin pressure at the wellhead), the 85% recovery factor indicates an abandonment pressure at a wellhead shutin pressure of about 65 psig. Many wells in those fields are now approaching the magic pressure of 65 psig, but there has been no movement to abondon the wells. Any sensible observer of practices in the gas industry knows that gas wells are abandoned when current income does not cover current costs of operations.

Historically, recoverable reserves for gas reservoirs have been determined by one or more of the following methods:

1. Volumetric methods with abandonment at an arbitrary shutin pressure

* Reference 7, chapter 7

2. Extrapolation of pressure-cumulative-production curves to an arbitrary abandonment pressure
3. Extrapolation of production rate-time decline curves to an abandonment rate of production
4. Type curve analysis based on theoretical producing rate-time curves

The author contends that recoverable reserves for gas can be determined only by a sensible forecast of production down to a limiting rate of production as determined by economic conditions. Therefore, methods for forecasting production from gas wells have been emphasized.

PRESSURE AND TEMPERATURE BASES FOR GAS MEASUREMENT

The natural gas industry has been plagued by an inability to agree upon a standard pressure base for gas measurement. It is not uncommon to find that production is reported to the regulatory body on one pressure base, reserves are kept on a second pressure base, and heating value is reported on a third pressure base. Further, the pressure base for reporting production varies from state to state. The province of Alberta, which has adopted the SI metric system, uses a measurement pressure base of 101.325 kPa and 15°C, which is equivalent to 14.696 psia and 59°F. Fortunately, the industry in the United States has not been plagued with a multitude of temperature bases for measurement purposes. The author has arbitrarily selected a pressure base of 14.65 psia and a temperature base of 60°F; all gas volumes used herein exist at these standard conditions unless stated otherwise. This has been done in the belief that most readers are required to report test results and production as standard cubic feet at 14.65 psia and 60°F.

NOMENCLATURE*

Symbol	*Description*	*Oil-field Units*
A	constant, equations 5–7 and 5–8	—
A	cross-sectional area	sq ft
A_1–A_8	constants, equation 2–11	—
A_x	availability function, equation 9–23	—

* Standard symbols proposed by SPE-AIME. Exceptions are made where strict conformance with the standard may be confusing.

Symbol	*Description*	*Oil-field Units*
b	gas constant for 1 lb of gas	cu ft, psia, °R^{-1}, lb^{-1}
B_g	formation volume factor for gas	—
B_1	constant, equation 5–7	—
B_2	constant, equation 5–8	—
cf	volume	cu ft
c'	coefficient, equations 4–4 and 4–5	—
c_g	compressibility of gas	psia^{-1}
c_t	total compressibility of a system	psia^{-1}
C	orifice-plate coefficient, problem 4–1	—
C	coefficient, equations 7–1, 7–3, 7–7, etc.	—
C (t)	coefficient, equation 7–2	—
d	diameter	in.
d	internal diameter of pipe	in.
d_1	outside diameter of inner pipe, annular space	in.
d_2	inside diameter of outer pipe, annular space	in.
D	internal diameter of pipe	ft
E	internal energy	ft-lb
f	coefficient of friction	—
$\sqrt{1/f}$	transmission factor	—
F	frictional resistance	ft-lb
°F	temperature, degrees Fahrenheit	°F
F_b	basic orifice factor	—
F_g	specific gravity factor	—
F_m	manometer factor	—
F_p	basic orifice prover factor	—
F_{pb}	pressure base factor	—
F_{pv}	supercompressibility factor	—
F_r	Reynolds number factor	—
F_r	factor defined by equation 5–11	—
F_t	flowing temperature factor	—
F_{tb}	temperature base factor	—
F_{tf}	flowing temperature factor	—

Symbol	*Description*	*Oil-field Units*
g	gravitational conversion factor (32.17)	ft/sec^2
G	volume of gas at standard conditions (14.65 psia and 60°F)	cu ft
G_i	gas initially in place	cu ft
G_p	gas produced	cu ft
h	thickness	ft
h_w	differential pressure	in. of water
H	elevation or difference in elevation	ft
I_n	factor used in chapter 5	—
ID	internal diameter	in.
k	permeability (chapter 6 only)	darcy
k	reservoir permeability	md
k	absolute roughness characteristic	in.
K	discharge coefficient	—
K	term in equation 2–1	—
ℓn	logarithm to base e	—
log	logarithm to base 10	—
L	length	ft
m	slope of straight-line, semilog coordinates	psia^2/cycle or psia^2/cp/cycle
m(p)	real gas potential	psia^2/cp
M	thousands	—
M	molecular weight	lb/lb-mole
M_a	average molecular weight air = 28.964	lb/lb-mole
M_i	molecular weight of component i	lb/lb-mole
M_g	molecular weight of gas	lb/lb-mole
Mcf	thousands of cubic feet	10^3 cf
MMcf	millions of cubic feet	10^6 cf
Mcfd	flow rate	10^3 cf/day
MMcfd	flow rate	10^6 cf/day

Symbol	*Description*	*Oil-field Units*
n	quantity of gas	lb-mole
n	coefficient in equations 7–1, 7–2, 7–3, etc.	—
N	mass rate of flow	lb/sec
N_{Re}	Reynolds number	—
OD	outside diameter	in.
p	pressure	psia
(p)	indicates preceding variable is a function of pressure	—
Δp^2	difference in pressures squared	$psia^2$
p_b	base pressure for gas measurement	psia
p_{cf}	pressure on casing at surface—no packer with flow-through tubing	psia
$\bar{p}_{cs}$	stabilized shutin pressure at surface on casing—no packer	psia
p_m	low base pressure for real gas potential	psia
p_m	measurement pressure	psia
p_{pc}	pseudocritical pressure	psia
p_{pr}	pseudoreduced pressure	—
$\bar{p}_R$	stabilized shutin reservoir pressure	psia
p_{tf}	surface pressure on tubing, flow through tubing	psia
p_{tfd}	deliverability pressure	psia
$\bar{p}_{ts}$	stabilized shutin pressure at surface on tubing, equation 7–7	psia
p_{wf}	bottom-hole flowing pressure	psia
p_{ws}	bottom-hole shutin pressure	psia
q	energy lost to surroundings	ft-lb
q	volumetric rate of flow	cfd
q_d	deliverability	Mcfd
q_D	dimensionless rate of flow	—
q_h	rate of flow	cf/hr
q_k	rate of flow	Mcfd
q_m	rate of flow	MMcfd
q_o	rate of flow for oil	bbl/day
q_{of}	open flow potential, equation 7–8	Mcfd

Symbol	*Description*	*Oil-field Units*
r	radius	in.
r	internal radius of pipe	in.
r_w	wellbore radius	ft
r'_w	effective wellbore radius	ft
R	gas constant	cu ft, psia, $°R^{-1}$, lb $mole^{-1}$
°R	°F + 460, degrees Rankine	°F + 460
R_g	gas to hydrocarbon liquid ratio	cu ft/bbl
s	factor in equations 5–19 and 5–20	—
s	van Everdingen skin effect, equation 8–23	—
S_w	water saturation	fraction
t	time (check usage)	days, years
(t)	preceding variable is function of time	—
t_D	dimensionless time	—
T	absolute temperature, °F + 460 = °R	°R
T_b	base temperature for gas measurement 60°F or (°F + 460) depending on usage	°F, °R
T_{pc}	pseudocritical temperature	°F + 460
T_{pr}	pseudoreduced temperature	—
u	velocity	ft/sec
$\bar{u}$	average velocity	ft/sec
v	specific volume	cu ft/lb
V	volume (chapter 6, cc)	cu ft
V_g	volume containing gas	cu ft
V_L	vapor volume equivalent of hydrocarbon liquid	cu ft/bbl
W	weight (chapter 6)	gm
W	energy supplied to system	ft-lb
W_f	energy loss caused by friction or irreversible energy loss	ft-lb
x	ratio of pressures $\bar{p}_R/p_{wf}$	—
x_f	fracture length	ft

Symbol	*Description*	*Oil-field Units*
X	see equation 2–21	—
X	length of pipe	miles
y_i	mole fraction of component i in gas mixture	—
Y	see equation 2–21	—
Y	expansion factor in gas measurement	—
z	compressibility factor	—
$\bar{z}$	compressibility factor at average conditions	—
γ_g	gas specific gravity (air = 1.000)	—
γ_{gf}	gas gravity of flowing fluid in a well	—
γ_o	specific gravity of oil referred to water	—
Δ	increment of the following variable	—
Δp^2	difference in pressures squared	$psia^2$
μ	viscosity	cp
$\bar{\mu}$	viscosity at average conditions	cp
ρ_a	density of dry air	lb/cu ft
ρ_g	density of gas	lb/cu ft
ρ_{pr}	pseudoreduced density, equation 2–10	—
ρ_L	density of liquid	lb/cu ft
ρ_1	density (chapter 2)	gm/cc
ϕ	porosity	fraction

Subscripts

a	air	—
b	standard or base conditions for gas measurement	—
b	bulk (chapter 6)	—

Symbol	*Description*	*Oil-field Units*
cs	casing, shutin (with pressure)	—
cf	casing, flowing (with pressure)	—
d	deliverability (with pressure)	—
e	external boundary	—
f	flowing	—
f	friction	—
f	fracture	—
g	gas	—
i	initial	—
i	i^{th} component	—
k	thousands	—
L	liquid	—
m	millions	—
m	measurement (with pressure)	—
ma	matrix	—
n	n^{th} term	—
o	oil	—
of	open flow	—
p	pore, see equation 6–1	—
p	prover	—
pc	pseudocritical	—
pr	pseudoreduced	—
r	related to radius, equation 5–11	—
R	reservoir	—
t	total system	—
tfd	deliverability (with pressure)	—
tfp	pipeline pressure	—

Symbol	*Description*	*Oil-field Units*
w	water	—
w	wellbore	—
wf	bottom-hole flowing (with pressure)	—
ws	bottom-hole shut in, (with pressure)	—
x	indicates a function of *x*, equation 9–23	—
1,2,3, etc.	indicates time or position	—

2 Properties of Natural Gas

A knowledge of the properties of natural gases is fundamental to applied natural gas engineering. The engineer concerned with the production, transportation, and sale of natural gas is interested in those properties that influence flow and sales value. These are pressure-volume-temperature relationships, viscosity, and composition of the gas. The heating value, or its quality for use as a fuel, controls the sale value—providing other variables remain constant.

This unit of measurement for natural gas at the present time is the cubic foot, which is defined as the quantity of gas contained in 1 cu ft of volume at a given base pressure and temperature. Unfortunately, there is no industry-wide standard base pressure; however, all gas volumes referred to as "standard" herein exist at a base pressure of 14.65 psia and a base temperature of 60°F unless otherwise specified. Normally, no distinction is made between the dry and the as-measured cubic foot since the terms are used interchangeably in natural gas measurement. Many contracts require a low water vapor content such as 7 lb/MMcf, and in most reservoir engineering work, the natural gas is assumed to be dry, or free of water vapor, unless gas hydrate problems are under consideration.

Natural gas is generally measured in producing operations at pressures several times greater than the base pressure and at temperatures substantially above the base temperature. Frequently, engineering studies require calculating the volume of a given quantity of gas at specified pressures and temperatures. These calculations are made by applying the gas laws.

GAS LAWS

A gas may be defined as a homogeneous fluid that fills any container in which it is placed. The physical laws that relate the effects of pressure and temperature on the volume of a given quantity of gas are well defined. The laws were first determined empirically in the seventeenth and eighteenth centuries and have been given the names of the persons credited with their discovery. Boyle's law states that, under conditions of constant temperatures, the volume of a gas varies inversely with the pressure. Difficulties with the formulation of a scientific concept of temperature delayed the discovery of the relationship between volume and temperature for nearly a century. Charles' or Gay-Lussac's law states that, under conditions of constant pressure, the volume of a gas varies directly as the absolute temperature.

When these laws are combined with Avogadro's hypothesis, which states that equal volumes of different ideal gases at the same pressure and temperature contain the same number of molecules, the ideal, or perfect, gas laws can be expressed by the equation:

$$pV/T = nR \tag{2–1}$$

Equation 2–1 relates pressure, volume, and temperature to the quantity of gas under consideration, n, and the gas constant, R. The development of the kinetic-molecular theory of gases gave the gas laws a theoretical basis.

Experimental work has shown that the ideal gas equation describes the pressure-volume-temperature behavior of natural gas only as a first approximation. For engineering purposes, the deviations from ideal behavior for natural gas have been correlated by means of the ratio pV/nRT, which is called the compressibility factor, z. For nonideal gases, the gas laws may be stated as follows:

$$pV = nzRT \tag{2–2}$$

where n is the quanitity of gas in moles and is defined as:

$$n = \frac{\text{weight of gas}}{\text{molecular weight of gas}} = \frac{\text{wt gas}}{\text{M}}$$

Gas Constant, *R*

The value of the constant R is dependent upon the units used in equation 2–2. For the units:

p = absolute pressure, psia
V = volume, cu ft
z = the compressibility factor, ratio
n = number of lb-moles

T = absolute temperature, (°F + 459.67), known as degrees Rankine or °R

the value of the gas constant, R, is:

$R = 10.732$ (cu ft, psia, $°R^{-1}$, lb-mole^{-1})

and:

$R = 1{,}545.4$ (cu ft, lb/sq ft, $°R^{-1}$, lb-mole^{-1})

For 1 lb of gas:

$$pV = bzT;\ b = \frac{R}{M} \tag{2–3}$$

Taking the average molecular weight of air as 28.964 and γ_g, the specific gravity of the gas referred to air, the molecular weight of the gas is as follows (see reference 5):

$$M = 28.964\ \gamma_g$$

then:

$$b = \frac{10.732}{28.964\ \gamma_g} = \frac{0.37053}{\gamma_g} \tag{2–4}$$

in units of cu ft, psia, $°R^{-1}$, and lb^{-1}, and:

$$b = \frac{(10.732)(144)}{28.964\ \gamma_g} = \frac{53.356}{\gamma_g} \tag{2–5}$$

in units of cu ft, lb/sq ft, $°R^{-1}$, and lb^{-1}

Compressibility Factor, *z*

In practice, the compressibility factor, z, is a variable, and its value depends upon the pressure, the temperature, and the composition of the gas. Plots of compressibility factors as a function of pressure for a nonideal gas at constant temperature take the form illustrated in figure 2–1.

Studies of the compressibility factors for natural gases of various compositions have shown that compressibility factors can be generalized with sufficient accuracies for most engineering purposes by introducing the concepts of pseudoreduced pressures and temperatures based on pseudocritical pressures and temperatures as follows:

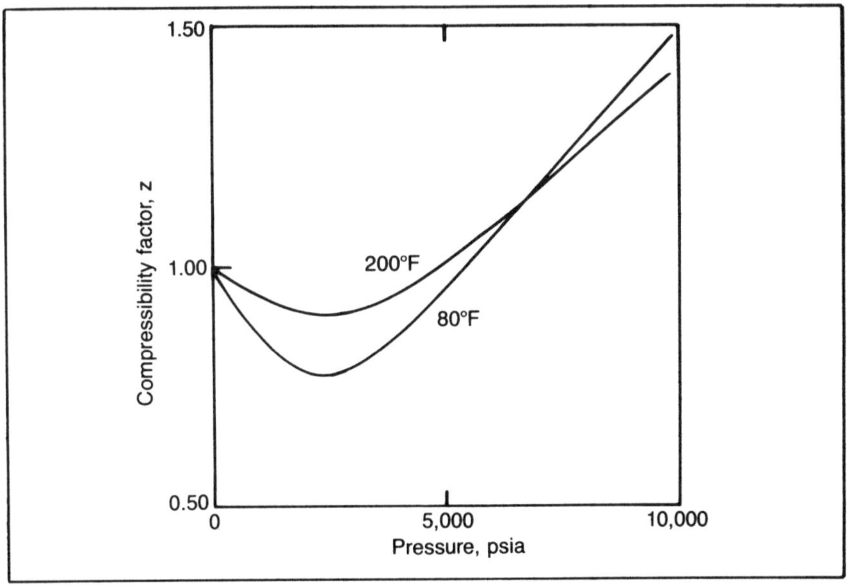

FIGURE 2–1 *Compressibility Factors for a Natural Gas as a Function of Pressure at Constant Temperature*

$$p_{pr} = p/p_{pc};\ T_{pr} = T/T_{pc}$$

where all pressures are in absolute units, psia, and temperatures are absolute, such as degrees Rankine (°F + 460).*

In natural gas engineering it has become common practice to use the compressibility chart of Standing and Katz for compressibility factors.[1] This chart was converted to a table by Poettmann and Carpenter and has been duplicated as table A–3 in appendix A.[2] Poettmann and Carpenter's table was expanded into a still larger table of compressibility factors by the Interstate Oil Compact Commission in its *Manual of Back-Pressure Testing of Gas Wells.*[3]

Table A–3 in appendix A gives values of compressibility factors for pseudoreduced pressures from 0.20 to 15.00 in increments of 0.05 and for values of pseudoreduced temperatures from 1.05 to 3.00 in convenient increments. A correlation from reference 3 between the specific gravity (air = 1.00) of the natural gas and pseudocritical pressures and temperatures is given as table A–1, and corrections to the pseudocritical properties for carbon dioxide,

* This approximation will be used for absolute temperature.

nitrogen, and hydrogen sulfide are given in table A–2 in appendix A. If the gas under consideration contains more than 5.0 vol% of carbon dioxide or hydrogen sulfide, the reader should consult the work of Robinson et al.[4] or the work of Wichert and Aziz that is summarized on pages 16–10 through 16–15 of reference 5.

The information in tables A–1 and A–2 can be approximated in the range of specific gravities from 0.55 to 0.90 by the following equations:

$$p_{pr} = 689.60 - 30.48\,\gamma_g + 4.35\,(\%\,CO_2) - 1.68\,(\%\,N_2) + 4.44\,(\%\,H_2S)$$

$$\begin{aligned} T_{pr} &= 152.89 + 339.29\,\gamma_g - 1.68\,(\%\,CO_2) \\ &\quad - 2.69\,(\%\,N_2) + 0.00\,(\%\,H_2S) \end{aligned} \tag{2–7}$$

Other means for calculating compressibility factors for natural gases have been proposed by Hall and Yarborough, Yarborough and Hall, Gopal, and Garb.[6–9] The equations proposed in references 6, 7, and 8 are readily adaptable to large computers, but Garb and Meehan and Lyons give equations and programs for small, hand-held programmable calculators.[10,11] Garb credited his information to the Energy Resources Conservation Board of Alberta, Canada Bulletin 75–34. The equations presented by Garb were published originally by Dranchuk, Purvis, and Robinson as follows:[12]

$$p_{pc} = 702.5 - 50\,\gamma_g \tag{2–8}$$

$$T_{pc} = 167 + 316.67\,\gamma_g \tag{2–9}$$

$$\rho_{pr} = 0.27\,p_{pr}/zT_{pr} \tag{2–10}$$

$$\begin{aligned} z = 1 &+ (A_1 + A_2/T_{pr} + A_3/T_{pr}^{\,3})\,\rho_{pr} \\ &+ (A_4 + A_5/T_{pr})\,\rho_{pr}^{\,2} \\ &+ (A_5 A_6\,\rho_{pr}^{\,5})\,/T_{pr} \\ &+ (A_7\,\rho_{pr}^{\,2}/T_{pr}^{\,3})(1 + A_8\,\rho_{pr}^{\,2})\exp(-A_8\,\rho_{pr}^{\,2}) \end{aligned} \tag{2–11}$$

where:

$A_1 = 0.31506237$
$A_2 = -1.0467099$
$A_3 = -0.57832729$

$$A_4 = 0.53530771$$
$$A_5 = -0.61232032$$
$$A_6 = -0.10488813$$
$$A_7 = 0.68157001$$
$$A_8 = 0.68446549$$

Comparison of Methods for Estimating Compressibility Factors

Thus, with no more than a hand-held programmable calculator, compressibility factors may be determined by interpolating data from table A–3 or from the tables in the Interstate Oil Compact Commission Manual or by calculations using methods given by Garb and Meehan and Lyons. A comparison is given in table 2–1 between compressibility factors from tables A–1, A–2, and A–3 and the methods of Garb, Hall and Yarborough, and Yarborough and Hall for a gas having a specific gravity of 0.641 and carbon dioxide and nitrogen contents of 2.0 and 3.0 vol %, respectively.

Although there are large differences between the sets of compressibility factors in table 2–1, the equations given by Hall and Yarborough fit the factors given in the table very closely. The important characteristic of a system for estimating compressibility factors for engineering purposes is that the system be continuous and smooth enough for material-balance calculations. For example, using one compressibility factor from the tables with another from the Garb system could introduce a large error into a material balance tabulation.

If equations 2–6 and 2–7 are used with equation 2–11 to calculate pseudocritical properties of a gas, very good agreement with the tables is

TABLE 2–1.

Compressibility Factors for a Natural Gas Interpolated from Tables A–1, A–2, and A–3 and Calculated from References 6, 7, and 9

$\gamma_g = 0.641$, $CO_2 = 2.0\%$. $N_2 = 3.0\%$

	Temperature								
	100°F			*150°F*			*200°F*		
Pressure (psia)	*Tables*	*Ref. 9*	*Ref. 6, 7*	*Tables*	*Ref. 9*	*Ref. 6, 7*	*Tables*	*Ref. 9*	*Ref. 6, 7*
500	0.936	0.929	0.935	0.954	0.947	0.953	0.967	0.960	0.966
1,000	0.878	0.864	0.878	0.913	0.902	0.913	0.938	0.928	0.938
1,500	0.834	0.813	0.833	0.881	0.867	0.882	0.917	0.904	0.917
2,000	0.807	0.785	0.805	0.861	0.848	0.863	0.903	0.892	0.905
2,500	0.798	0.781	0.799	0.852	0.844	0.858	0.898	0.891	0.902
3,000	0.812	0.797	0.811	0.859	0.855	0.867	0.906	0.900	0.910

obtained at pressures up to 2,000 psia, but the divergence becomes large at higher pressures.

Supercompressibility Factor, F_{pv}

The term used to express the deviation of a gas from ideal behavior appears in gas measurement equations as the square root of the reciprocal of the compressibility factor, z. This term has been called the supercompressibility factor, F_{pv}:

$$F_{pv} = \sqrt{1/z} \tag{2–12}$$

DENSITY AND SPECIFIC VOLUME, ρ AND *v*

Density, ρ, and specific volume, v, may be calculated using equations 2–2 and 2–3 when 1 lb of gas is considered.

$$pv = bzT \tag{2–2}$$

$$b = \frac{0.37053}{\gamma_g} \tag{2–4}$$

$$\frac{1}{\rho} = v = \frac{0.37053\ zT}{p\ \gamma_g} \tag{2–13}$$

$$\rho = \frac{1}{v} = 2.6988 \frac{p\ \gamma_g}{zT} \tag{2–14}$$

where the units are lb/cu ft for density and cu ft/lb for specific volume.

Example 2–1. Determine the density and specific volume of a gas (γ_g = 0.641, CO_2 = 2.0%, and N_2 = 3.0%) at 2,765 psia and 179°F.

$$p_{pc} = 670 + 8 - 5 = 673$$

$$p_{pr} = 2{,}765/673 = 4.11$$

$$T_{pc} = 372 - 3 - 9 = 360$$

$$T_{pr} = (460 + 179)/360 = 1.78$$

	T_{pr}	
p_{pr}	*1.70*	*1.80*
4.10	0.857	0.893
4.15	0.857	0.894

$$z \text{ at } p_{pr} = 4.11 \text{ and at } T_{pr} = 1.78 = 0.886$$

$$\rho = \frac{(2.6988)(2{,}765)(0.641)}{(0.886)(639)} = 8.45 \text{ lb/cu ft}$$

$$v = 1/8.45 = 0.118 \text{ cu ft/lb}$$

COMPRESSIBILITY, c_g, FOR GASES

Many reservoir engineering equations contain terms defined as the compressibilities of various fluids. The isothermal compressibility of a substance is defined as the change in volume per unit volume for a unit change in pressure, or:

$$c_g = -\frac{1}{V}\left(\frac{\partial V}{\partial p}\right)_T$$

The general gas equation (equation 2–2) can be differentiated and solved for compressibility with the following result:

$$c_g = \frac{1}{p} - \frac{1}{z}\left(\frac{\partial z}{\partial p}\right)_T \qquad (2\text{–}15)$$

Inspection of equation 2–15 shows that the compressibility for an ideal gas would be the reciprocal of the absolute pressure. The value of $\partial z/\partial p$ can be determined graphically from curves showing the compressibility factor plotted against pressure as illustrated in figure 2–2.

Note: Compressibility, c_g, must not be confused with the compressibility factor, z.

Example 2–2. Refer to example 2–1 and calculate the compressibility of the same gas at 1,800 psia and 179°F. Since the compressibility calculation requires $(\partial z/\partial T)_T$ at the required pressure and temperature, it is necessary to construct a curve (see figure 2–2) showing the relationship between the compressibility factor and pressure.

From figure 2–2:

$$\left(\frac{\partial z}{\partial p}\right)_T = \left(\frac{\Delta z}{\Delta p}\right)_T = \frac{0.860 - 0.872}{1{,}900 - 1{,}600} = -4.0 \times 10^{-5},\ \text{psia}^{-1}$$

From equation 2–15:

$$c_g = \frac{1}{1{,}800} - \left[\frac{1}{0.864}\right]\left(-4.0 \times 10^{-5}\right)$$

$$= 5.56 \times 10^{-4} + 0.46 \times 10^{-4} = 6.02 \times 10^{-4},\ \text{psia}$$

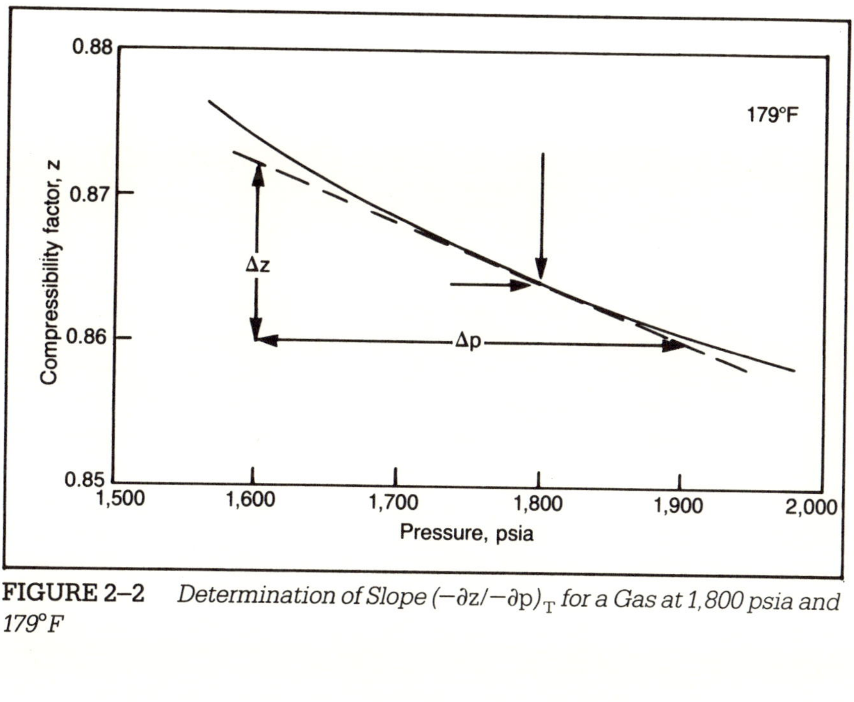

FIGURE 2–2 *Determination of Slope* $(-\partial z/-\partial p)_T$ *for a Gas at 1,800 psia and 179°F*

FORMATION VOLUME FACTOR, B_g

The reservoir volume, B_g, occupied by a unit volume of gas at standard conditions can be calculated from equation 2–2 with the following result:

$$B_g = \frac{p_b \, z \, T}{p \, z_b \, T_b} \tag{2–16}$$

where the subscript b denotes standard or base conditions of pressure and temperature. Usually, the compressibility factor at standard conditions is assumed to be 1.000.

Example 2–3. Calculate the formation factor for the gas in example 2–2 at 1,800 psia and 179°F for a pressure base of 14.65 psia and 60°F.

$$B_g = \frac{(14.65)\,(0.864)\,(460 + 179)}{(1{,}800)\,(1.000)\,(520)} = 0.00864$$

Thus, 1 cu ft of gas at 14.65 psia and 60°F would occupy 0.00864 cu ft at 1,800 psia and 179°F. **Note:** 460 was used to convert °F to absolute temperature instead of the value of 459.67 given in the discussion of the gas constant, R. This approximation will be used in all examples.

MOLE FRACTION, VOLUME FRACTION, AND WEIGHT FRACTION

For many engineering purposes, the approximation that natural gas and air are ideal gases at atmospheric pressure and ordinary ambient temperatures of about 60°F or at the commonly used standard pressure and temperature conditions can be made. Actually, the compressibility factors for natural gases range from 0.997 to 0.998 at atmospheric pressure and temperature of 60°F, and the compressibility factor for dry air is about 0.9996 at atmospheric pressure and 60°F. Using the approximation that the gases are ideal gases, the number of moles of gas, n, is directly proportional to the volume, and the mole fraction of a constituent in the gas is equivalent to the volume fraction.

To convert mole fraction to weight fraction, one must multiply the mole fraction of each component in the gas by the molecular weight of the component, add the weights for the individual components, and divide the weight of each component by the total. Thus:

$$\text{weight fraction} = \frac{(y_i \times M_i)}{\Sigma\,(y_i \times M_i)}$$

where y is the mole fraction of the ith component in the mixture. Weight fractions can be converted to mole fractions by the inverse operation. Molecular weights for the components commonly found in natural gas are given in many reference books but are included in table 2–3.

APPARENT MOLECULAR WEIGHT

A gas mixture behaves as if its molecular weight is the same as a pure gas. The weight of 1 mole of a gas mixture and, thus, its average molecular weight is:

$$\text{average molecular weight} = \Sigma\,(y_i \times M_i)$$

Example 2–4. A gas consists of 17 lb of methane, 2 lb of ethane, and 1 lb of propane. Calculate the compressibility factor at 1,000 psia and 100°F.

Component	*(1) Weight (lb)*	*(2) Weight Fraction*	*(3) Molecular Weight*	*(1) ÷ (3) (Moles)*
Methane	17	0.85	16.043	1.0597
Ethane	2	0.10	30.070	0.0665
Propane	1	0.05	44.097	0.0227
Total	20	1.00		1.1489

$$\text{Average molecular weight} = \frac{\text{weight}}{\text{moles}} = \frac{20}{1.1489} = 17.41$$

$$\text{Specific gravity} = \frac{\text{average molecular weight}}{\text{average molecular weight air}} = \frac{17.41}{28.964} = 0.601$$

From table A–1, pseudocritical pressure = 671 psia
pseudocritical temperature = 358°F absolute

$$\text{Pseudoreduced pressure} = \frac{1{,}000}{671} = 1.49$$

$$\text{Pseudoreduced temperature} = \frac{(460 + 100)}{358} = 1.56$$

From table A–3, the compressibility factor $z = 0.878$

Example 2–5. A gas reservoir has an area of 4 sq mi and a net pay thickness of 10 ft. The porosity is 10%, connate water is 25%, reservoir temperature is 85°F, and reservoir pressure is 700 psia. Calculate the volume of gas in place at 14.65 psia and 60°F and the gas remaining in place when the reservoir pressure has been reduced to 50 psia. The gas composition is given as weight fractions in the following table:

Component	*Weight Fraction*	*Molecular Weight*	*Moles/lb*	*Mole Fraction*
Methane (C_1)	0.94	16.043	0.05859	0.973
Ethane (C_2)	0.03	30.070	0.00100	0.017
Propane (C_3)	0.02	44.097	0.00045	0.007
n-Butane (n-C_4)	0.01	58.124	0.00017	0.003
Total	1.00		0.06021	1.000

$$\text{Molecular weight} = \frac{\text{weight}}{\text{moles}} = \frac{1.000}{0.06021} = 16.61$$

$$\text{Specific gravity} = \gamma_g = \frac{M_g}{M_a} = \frac{16.61}{28.964} = 0.573$$

$p_{pc} = 672$ psia
$T_{pc} = 347$°F absolute

At $p = 700$ psia and 85°F:

$p_{pr} = 700/672 = 1.04$
$T_{pr} = (85 + 460)/347 = 1.57$
$z = 0.914$

At $p = 50$ psia and 85°F:

$p_{pr} = 50/672 = 0.07$
$T_{pr} = (85 + 460)/347 = 1.57$
$z = 0.988$

Volume holding gas in reservoir

= (area) (thickness) (porosity) (1 − water saturation)

$$V = 4(5{,}280)^2 (10) (0.10) (1 - 0.25) = 83.6 \times 10^6, \text{ cu ft}$$

where V is the actual volume of gas in reservoir at 700 psia and 85°F:

$$\frac{p_b G}{z_b T_b} = \frac{pV}{zT}$$

$$\frac{(14.65)\ G}{(1.00)\ (60 + 460)} = \frac{(700)\ (83.6 \times 10^6)}{(0.914)\ (85 + 460)}$$

$$G = 4{,}170 \times 10^6 \text{ cu ft at 14.65 psia and 60°F}$$

Volume of gas, G, in reservoir at 50 psia and 85°F:

$$\frac{(14.65)G}{(1.00)\ (60 + 460)} = \frac{(50)\ (83.6 \times 10^6)}{(0.988)\ (85 + 460)}$$

$$G = 276 \times 10^6 \text{ cu ft at 14.65 psia and 60°F}$$

Gas removed from the reservoir to reduce pressure from 700 to 50 psia is:

$$4{,}170 \times 10^6 - 276 \times 10^6 = 3{,}894 \times 10^6 \text{ cu ft at 14.65 psia and 60°F}$$

VOLUME OF A POUND-MOLE

Occasionally in natural gas engineering computations, it is convenient to know the volume of a pound-mole of an ideal gas at the various standard pressure bases for gas measurement. Taking the volume of 1 lb-mole of an ideal gas as 379.49 cu ft at 14.696 psia and 60°F the volumes at other commonly used pressure bases are given in table 2–2 (see reference 5). If the compressibility factor is known for the actual gas, the volumes given in table 2–2 should be multiplied by the compressibility factor for the gas at the base pressure and temperature.

TABLE 2–2.
Volumes of 1 lb-mole of an Ideal Gas at Various Pressure Bases for Gas Measurement and at 60°F (based on GPSA Engineering Data Book*)*

Pressure Base, psia	*Volume at 60°F, cu ft*
15.025	371.18
14.73	378.61
14.696	379.49
14.65	380.68
14.4	387.29

SPECIFIC GRAVITY, γ_g, OF NATURAL GAS

The specific gravity of natural gas usually refers to air and may be defined as:

$$\text{specific gravity} = \gamma_g = \frac{M_g}{M_a} = \frac{M_g}{28.964} \text{ at } 60^\circ\text{F}$$

In field practice, the specific gravity of a gas is readily determined using recording instruments or a gas-gravity balance. In many cases, the specific gravity of the gas is known when the actual composition is not. If the composition is known, the specific gravity can be calculated from the composition as illustrated in the following section, "Gas Calorimetry."

Occasionally, the only information available may be a bottom-hole pressure survey run with a subsurface pressure gauge. The pressure gradient information can be used with equation 5–20 in chapter 5 to estimate the specific gravity of the gas.

Gas Calorimetry

Since the heating value of natural gas determines its worth as a fuel, the production engineer must have an elementary knowledge of gas calorimetry. The heat content of a gas is the principal measure of its utility in heating service. The following definitions are used in discussing the heating value of natural gas.

A British thermal unit, Btu, is a unit of measurement for the quantity of heat that must be added to 1 lb avoirdupois of pure water to raise its temperature from 58.5 to 59.5°F under standard pressure.

Total or gross heating value or total or gross calorific value is the number of Btu's evolved by the complete combustion at constant pressure of 1 scf of

TABLE 2–3.
Calculation of Molecular Weight, Heating Value, and Specific Gravity from Gas Composition

Component	(1) Mole Fraction	(2) Molecular Weight	(3) = (1) × (2)	(4) Gross Heating Value	(5) = (1) × (4)
Helium (He)	0.0054	4.003	0.0216	0	0
Carbon dioxide (CO_2)	0.0003	44.010	0.0132	0	0
Nitrogen (N_2)	0.1297	28.013	3.6333	0	0
Methane (C_1)	0.8283	16.043	13.2884	1,009.7	836.3
Ethane (C_2)	0.0230	30.070	0.6916	1,768.8	40.7
Propane (C_3)	0.0088	44.097	0.3881	2,517.4	22.2
Isobutane (iC_4)	0.0013	58.124	0.0756	3,252.7	4.2
Butane (nC_4)	0.0021	58.124	0.1221	3,262.1	6.9
Isopentane (iC_5)	0.0004	72.151	0.0289	4,000.3	1.6
Pentane (nC_5)	0.0003	72.151	0.0216	4,009.5	1.2
Hexanes (C_6)	0.0002	86.178	0.0172	4,756.0*	1.0
Heptanes plus (C_7+)	0.0002	120.000*	0.0240	7,000.0*	1.4
Total	1.0000		18.3256		915.5
Pentanes plus (C_{5+})		85.0*		5,000.0*	
Hexanes plus (C_{6+})		100.0*		6,000.0*	

$$\text{Specific gravity} = \frac{\text{average molecular weight gas}}{\text{average molecular weight air}} = \frac{18.3256}{28.964} = 0.633$$

Gross heating value = 916 Btu/cu ft (dry basis)
916 × 0.9826 = 900 Btu/cu ft (wet basis)

* = Estimated

Note: Molecular weights and heating values taken from reference 5.

gas with air, with the temperature of the gas, air, and products of combustion at 60°F and all water formed by the combustion reaction condensing to the liquid state.

Net heating value or net calorific value is the number of Btu's evolved by the complete combustion at constant pressure of 1 scf of gas with air, with the temperature of the gas, air, and products of combustion at 60°F and all water formed by the combustion reaction remaining in the vapor state. The net calorific value is the total calorific value minus the latent heat of vaporization at the standard temperature of the water formed by the combustion reaction.

Standard cubic foot (scf) of gas is the quantity of any gas that, at standard pressure and temperature, will fill a space of 1 cu ft (dry basis). (On a net basis, the cubic foot of gas is in equilibrium with liquid water.)

Standard pressure for gas calorimetry usually is an absolute pressure of 14.696 psia (1 atm) but sometimes is defined as the absolute pressure of a column of pure mercury 30 in. high at 32°F under standard gravity of 32.174 ft/sec^2 or 14.735 psia.

Standard temperature is 60°F.

Note: One standard cubic foot of gas in equilibrium with water contains enough water vapor in the volume ratio that the partial pressure of the water bears to the standard pressure. Thus, at 60°F:

vapor pressure of water at 60°F = 0.2563 psia[13]

water fraction = 0.2563/14.696 = 0.0174

gas fraction = (14.696 − 0.2563)/14.696 = 0.9826

An example of the calculation of the gross heating value from the composition and the gross heating value of the pure components is shown in table 2–3. The steps in calculating the specific gravity of the gas from the molecular weights of the pure components and from the molecular weight of air is also shown in table 2–3.

VISCOSITY OF NATURAL GASES

The viscosity of a Newtonian fluid is defined as the ratio of the shear force per unit area to the local velocity gradient. A convenient chart for determining the viscosities of natural gases is duplicated in figure 2–3. Other charts for estimating viscosity have been presented by Katz et al.[14] A convenient set of empirical equations for calculating the viscosity, μ, of natural gases in centipoises has been published by Lee et al.[15] Adjusting these equations to give the viscosity in centipoises results in:

$$\mu = 10^{-4} K \exp (X \rho_1{}^{Y}) \qquad (2\text{–}17)$$

where:

$$K = \frac{(9.4 + 0.02M)\, T^{1.5}}{209 + 19M + T}$$

$$X = 3.5 + 986/T + 0.01M$$

$$Y = 2.4 - 0.2X$$

and:

ρ_1 = density in gm/cc

$$\rho_1 = (1.4926 \times 10^{-3}) \frac{pM}{zT} \qquad (2\text{–}18)$$

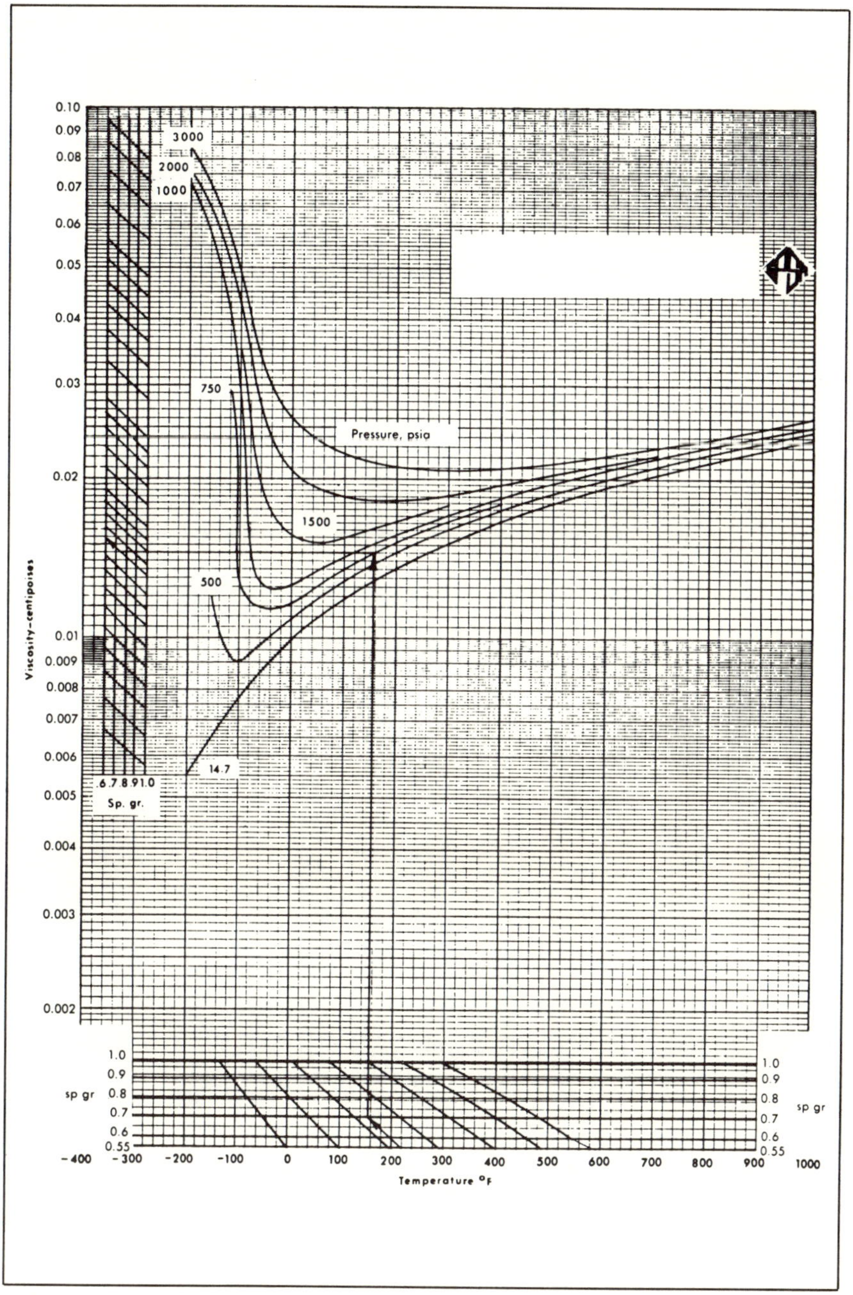

FIGURE 2–3 *Hydrocarbon Gas Viscosity (source:* GPSA Engineering Data Book*)*

Lee et al. measured experimental viscosities for four natural gases from 100 to 340°F and pressures from 100 to 8,000 psia. They found that equation 2–17 reproduced the experimental data with a standard deviation of ± 2.7% and a maximum deviation of 9.0%.

Compressibility factors from tables A–1, A–2, and A–3 were used to calculate densities using equation 2–18 which, in turn, were used in equation 2–17 to calculate the viscosities at 220°F of natural gas No. 2 from reference 15. These viscosities are shown in table 2–4 along with viscosities taken from figure 2–3 and the experimentally measured viscosities.

The agreement between experimental and calculated viscosities and those from figure 2–3 are believed to be very good. Values from either source are probably accurate enough for most engineering calculations. The viscosities of natural gases are used in calculating Reynolds numbers and in flow equations in calculations with the permeability of the reservoir rock.

TABLE 2–4.
Viscosities at 220°F for Natural Gas No. 2 in Reference 15

Pressure (psia)	Viscosity at 220°F, cp Experimental	Calculated	Figure 2–3
700	0.01426	0.01408	—
1,000	0.01486	0.01465	0.0152
1,500	0.01613	0.01588	0.0165
2,000	0.01772	0.01745	0.0186
3,000	0.02157	0.02133	0.0222
4,000	0.02555	0.02522	—
5,000	0.02950	0.02894	—
6,000	0.03315	0.03236	—

CALCULATION OF THE GAS GRAVITY OF THE FLOWING FLUID IN A WELL, γ_{gf}

In calculating subsurface pressures from measured wellhead pressures, one must know the specific gravity (air = 1.000) of the flowing fluid, γ_{gf}, in order to account for the weight of the column of gas. If the well produces dry gas (no hydrocarbon liquids), the specific gravity can be measured directly or calculated from the analysis. For wells equipped with a separator for removing liquid from the gas stream and a stock tank for storing the liquid, one must calculate the specific gravity of the flowing fluid from that of the separator gas, the gas-to-hydrocarbon liquid ratio, and the specific gravity of the hydrocarbon liquid.

In test example 1 in chapter 7, the specific gravity of the gas from the separator is 0.625 and the gas-to-hydrocarbon liquid ratio is 193,000 cu ft/bbl. Since the hydrocarbon liquid has a gravity of 50.2° API in the stock tank and the gas-to-liquid ratio is relatively high, the hydrocarbon liquid probably condensed out of the flowing fluid as it moved up the wellbore. At any rate, as it moved up the wellbore, the flowing fluid lifted the hydrocarbon liquid, either as a vapor or as a liquid; therefore, it is necessary to calculate the specific gravity, γ_{gf}, of the flowing fluid from the gas-to-hydrocarbon liquid ratio, R_g, in cu ft/bbl from the specific gravity of the separator gas referred to air, γ_g, from the specific gravity of the hydrocarbon liquid, γ_o, referred to water, and from the vapor volume equivalent of the hydrocarbon liquid, V_L, in cu ft/bbl. The gas volume and weight associated with 1 bbl of hydrocarbon liquid is:

$$\text{Volume, cu ft (14.65 psia and 60°F)} = R_g + V_L$$

$$\text{Weight, lb} = R_g \gamma_g \rho_a + 5.6146 \rho_L$$

where:

ρ_a = density, lb/cu ft at 14.65 psia and 60°F of dry air
ρ_L = density, lb/cu ft at 60° of the separator liquid

Here, one customarily assumes that the density of the hydrocarbon liquid in the separator is equal to that of the liquid in the stock tank; this is done to save computations.

$$\rho_a = \frac{\text{average molecular weight of air}}{\text{cu ft lb-mole (14.65 and 60°F)}} = \frac{28.964}{380.68} = 0.07608$$

$$\rho_{gf} = \frac{R_g \gamma_g (0.07608) + 5.6146 \rho_L}{R_g + V_L} \tag{2–19}$$

and:

$$\gamma_{gf} = \frac{\text{density of flowing fluid}}{\text{density of air}}$$

since:

$$\rho_L = \gamma_o \text{ (density of water)}$$

$$\text{density of water} = (7.4805 \text{ gal/cu ft}) \times (8.3372 \text{ lb/gal at } 60°\text{F}) = 62.3664 \text{ lb/cu ft}$$

From equation 2–19:

$$\gamma_{gf} = \frac{R_g \gamma_g + 4603 \gamma_o}{R_g + V_L} \tag{2–20}$$

The specific gravity and the approximate vapor volume of the hydrocarbon liquid can be calculated from the API gravity by:

$$\gamma_o = \frac{141.5}{131.5 + (\text{API})} \tag{2–21}$$

and:

$$V_L = 369 + 5\,(\text{API}) + 0.04\,(\text{API})^2 \tag{2–22}$$

where (API) = degrees API.

Except for a slight difference in the constant, equation 2–20 is the same as an equation in the IOCC Manual, and equation 2–22 is given in that manual.[16] Equation 2–21 is given in many textbooks. Thus, for test example 1 in chapter 7:

$$V_L = 369 + 5(50.2) + 0.04(50.2)^2 = 721$$
$$\gamma_o = 141.5/(131.5 + 50.2) = 0.7788$$

and:

$$\gamma_{gf} = \frac{193{,}000(0.625) + 4{,}603(0.7788)}{193{,}000 + 721} = 0.641$$

Note: Equation 2–20 is approximate because it does not take into account the hydrocarbon vapor that escapes from the liquid in a separator-stock tank system when the hydrocarbon liquid moves from the pressure and temperature of the separator to that of the stock tank.

PROBLEMS

1. A tank with a volume of 40 cu ft is filled with a carbon dioxide and air mixture. The pressure within the tank is 30.0 psia at 70°F. It is known that 2.0 lb of carbon dioxide was placed in the tank. Assume that air is 80% nitrogen and 20% oxygen and use the ideal gas laws. Calculate:

   a. Weight of air in the tank
   b. Weight percent of each constituent
   c. Volume percent of each constituent
   d. Mole fraction of each constituent
   e. Partial pressure of each constituent
   f. Average molecular weight of the mixture
   g. Density of the mixture at 20 psia and 70°F, at 14.7 psia and 60°F, and at 14.7 psia and 32°F
   h. Specific gravity of the mixture (air = 1.000)

2. Given the following gas:

Component	*Volume (%)*
Methane	58.7
Ethane	16.5
Propane	9.9
Butanes	5.0
C_{5+}	3.5
Hydrogen sulfide	6.4
	100.0

Calculate:

   a. Average molecular weight
   b. Density at 32°F and 14.7 psia
   c. Density at 0°C and 760 mm mercury
   d. Specific gravity and gross and net heating value

3. Construct a chart of compressibility factors plotted against pressure for pressures at 500 psi intervals to 3,000 psia and at temperatures of 100, 150, and 200°F for natural gas with the composition given in table 2–3.

4. Repeat problem 3 for a natural gas with the specific gravity $\gamma_g = 0.641$, $CO_2 = 2.0\%$, and nitrogen = 3.0%.

5. Using the results of problem 4, calculate the compressibility and the formation volume factors at 2,700 psia and 200°F.

6. Calculate the viscosity of the natural gas in problem 4 at 2,700 psia and 200°F using equation 2–17; compare the answer with the value read from figure 2–3.

REFERENCES

1. M.B. Standing and D.L. Katz. "Density of Natural Gases." *Trans., AIME,* 146 (1942), p. 140.
2. F.H. Poettmann and P.G. Carpenter. "The Multiphase Flow of Gas, Oil, and Water through Vertical Flow Strings with Application to the Design of Gas-Lift Installations." *API Drilling and Production Practice,* (1952), pp. 279–91.
3. Interstate Oil Compact Commission. *Manual of Back-Pressure Testing of Gas Wells.* Oklahoma City: IOCC, 1962.
4. D.B. Robinson, C.A. Macrygeorgos, and G.W. Govier. "The Volumetric Behavior of Natural Gases Containing Hydrogen Sulfide and Carbon Dioxide." *Trans., AIME,* 219 (1960), p. 54.
5. Gas Processors Suppliers Association. *Engineering Data Book,* 9th ed., Tulsa (1972), 3rd rev., 1977.
6. K.R. Hall and L. Yarborough. "A New Equation of State for Z-Factor Calculations." *OGJ,* 71, No. 25 (June 18, 1973), p. 82.
7. L. Yarborough and K.R. Hall. "How to Solve Equation of State for Z-Factors." *OGJ,* 72 (February 18, 1974), p. 86.
8. V.N. Gopal. "Gas Z-Factors." *OGJ,* 75 (August 8, 1977), p. 58.
9. F.A. Garb. "Property Evaluation with Hand-held Calculators, Part II—Moving Averages and Gas Deviation Factor." *Petrol. Eng. Int.,* 50 (December 1978), p. 56.
10. Garb, 1978.
11. D.N. Meehan and W.K. Lyons. "Calculations Programmable for Gas Compressibility." *OGJ* (October 8, 1979), pp. 74–78.
12. P.M. Dranchuk, R.A. Purvis, and D.B. Robinson. "Computer Calculation of Natural Gas Compressibility Factors Using

the Standing and Katz Correlations." *Inst. of Pet. Tech.*, IP-74-008.

13. J.H. Keenan and F.G. Keyes. *Thermodynamic Properties of Steam*. 35th printing, 1st ed., New York: John Wiley & Sons Inc.
14. Donald L. Katz, et al. *Handbook of Natural Gas Engineering*. New York: McGraw-Hill, 1959.
15. A.L. Lee, M.H. Gonzales, and B.E. Eakin. "The Viscosity of Natural Gases." *Trans., AIME*, 237 (1966), p. 997.
16. Interstate Oil Compact Commission, 1962.

3 Application of Gas Laws to Reservoir Engineering

The general gas laws and the concept that a gas reservoir is an idealized container form the basis for estimating the volume of gas contained in reservoirs. The extent to which a gas reservoir departs from the behavior of an idealized container depends upon the saturation and mobility of other fluids, the porosity and especially the permeability of the reservoir rock, and the capacity of the rock and interstitial liquids to permit the flow of the gas to production wells. In considering the ideal gas reservoir, the following assumptions will be made:

1. The volume of pore space in the reservoir containing gas remains constant throughout the productive life of the reservoir (i.e., no significant water or oil migration occurs, and there is no appreciable compressibility of the rock matrix).
2. Shutin pressures are representative of actual reservoir pressures.
3. Gas production data include all gas produced from the reservoir.
4. No gas is added to the reservoir by the evolution of solution gas from oil or water.

VOLUMETRIC DETERMINATION OF GAS IN PLACE

For a given quantity of gas, the terms, n and R, may be eliminated from the general gas equation (equation 2–2) to give the following:

$$\frac{pV_g}{zT} = \frac{p_b G}{T_b z_b} \tag{3–1}$$

where:

V_g = volume of space-holding gas, cu ft
G = volume of gas in the reservoir at standard conditions, cu ft
b = standard conditions for gas measurement

If 1 acre-ft of reservoir rock is under consideration, the volume in cubic feet, V_g, capable of holding gas when the porosity, ϕ, and the water saturation, S_w, are known, is:

$$V_g = 43{,}560\ \phi(1 - S_w) \tag{3–2}$$

Combining equations 3–1 and 3–2 and solving for the gas in place, G (expressed in cubic feet at standard conditions), the result is:

$$G = 43{,}560\ \phi(1 - S_w)pT_b/(zp_b T) \tag{3–3}$$

where the approximation has been made that z_b is unity. At pressure and temperature bases of 14.65 psia and 60°F, equation 3–3 becomes:

$$G \text{ (per acre-foot)} = 1{,}546{,}000\ \phi(1 - S_w)p/(zT) \tag{3–4}$$

where G is expressed in standard cubic feet. If one follows the natural gas industry convention by expressing volume in thousands of cubic feet (Mcf), equation 3–4 becomes:

$$G_k = Mcf = 1{,}546\ \phi(1 - S_w)p/(zT) \tag{3–5}$$

Equation 3–5 is used in volumetric estimates of the quantity of gas in place per acre-foot of reservoir.

Example 3–1. The gas used in example 2–6 had a specific gravity of 0.573, $p_{pc} = 672$, and $T_{pc} = 347$. Calculate the gas in place per acre-foot when the reservoir pressure was 600 psia at the reservoir temperature of 85°F. Porosity = 10% and S_w = 25%.

$p_{pr} = 600/672 = 0.89$
$T_{pr} = (460 + 85)/347 = 1.57$
$z = 0.926$

$$G_k = Mcf = \frac{(1{,}546)\ (0.10)\ (0.75)\ (600)}{(0.926)\ (85 + 460)} = 137.9$$

or 137.9 Mcf/acre-ft at standard pressure and temperature of 14.65 psia and 60°F, respectively. At a standard pressure of 14.73 psia, the volume would be:

$$\frac{(137.9)\ (14.65)}{(14.73)} = 137.2 \text{ Mcf/acre-ft.}$$

MATERIAL BALANCE

Pressure-production methods for estimating gas reserves consist of various types of plots of reservoir pressure corrected for the compressibility of the gas against cumulative production and extrapolating the curve to an abandonment pressure to arrive at the recoverable gas reserve. These methods and their variations are based on a material balance concept and the general gas law. The direct application of material balance principles to determine the quantity of gas in place involves the assumptions that the reservoir is closed and that there is no water drive or intrusion of oil. The material balance equation is:

$$G_p = G_i - G \tag{3–6}$$

This states that the quantity of gas produced is equal to the initial quantity in the reservoir minus the gas remaining in the reservoir at the time under consideration. The applicable gas laws for a reservoir that has a volume of space, V, containing gas are:

$$\frac{p_i V}{z_i T} = \frac{p_b G_i}{T_b} \text{ and } \frac{pV}{zT} = \frac{p_b G}{T_b}$$

where z_b is assumed to be 1.000. Substituting the above into equation 3–6 yields the equations that are commonly used in work with pressure-production curves for gas wells. The resulting equation is:

$$\frac{p}{z} = \frac{p_i}{z_i} - \frac{p_b T}{V T_b} (G_p) \tag{3–7}$$

Equation 3–7 represents the form, $a = mx + b$, and is a straight line when plotted on cartesian coordinate paper. This equation is the basis for plotting *p/z* against cumulative production on cartesian coordinates. If the reservoir meets with the requirements set out at the start of chapter 3, the result is a straight line. Further rearrangement of equation 3–7 results in the following:

$$\log(p_i/z_i - p/z) = \log(G_p) + \log(p_b T/VT_b) \tag{3–8}$$

This is an equation of the form ($\log y = \log x + \log b$), and a plot of ($p_i/z_i - p/z$) against cumulative production on log-log coordinates results in a straight line with a slope of unity. It forms a 45° angle with either axis. The term on the right in equations 3–7 and 3–8 is constant for any given reservoir if it is a closed reservoir. This approach was discussed in detail by Gruy and Crichton (see reference 5).

Example 3–2. The following pressures and cumulative production are for a gas well with a reservoir temperature of 112°F. Gas composition is 0% carbon dioxide, 3.0% nitrogen, and 0% hydrogen sulfide. Specific gravity is 0.666. Prepare plots of the pressure-production relationships on cartesian and log-log coordinates and estimate the original gas in place.

	Given		*Calculated*			
Time	*Pressure (psia)*	*Cumulative Production (MMcf)*	p_{pr}	z	p/z	$(p_i/z_i - p/z)$
0	1,556	0	2.34	0.821	1,895	0
1	1,543	19	2.32	0.821	1,879	16
2	1,538	45	2.32	0.821	1,873	22
3	1,504	108	2.27	0.824	1,825	70
4	1,443	214	2.17	0.830	1,739	156
5	1,432	275	2.16	0.830	1,725	170
6	1,423	319	2.14	0.831	1,712	183
7	1,360	427	2.05	0.836	1,627	268
8	1,274	522	1.92	0.843	1,511	384

$$p_{pc} = 669 + 0 - 5 + 0 = 664$$

$$T_{pc} = 380 - 0 - 9 - 0 = 371$$

$$T_{pr} = (112 + 460)/371 = 1.54$$

The original gas in place derived from figure 3–1 is 2,780 MMcf and from figure 3–2, 2,950 MMcf.

PROBLEMS IN USING MATERIAL BALANCE METHODS

Material balance methods for estimating gas reserves are widely used and, in general, are much preferred by the author over volumetric methods where conditions are favorable and production information is available. Although plotting pressure divided by compressibility, p/z, against cumulative production on cartesian coordinates is generally accepted, the log-log plot is of historical interest, and the method has serious disadvantages—and some compensating advantages—when applied to new wells in new reservoirs. Regardless of the type of plot used to interpret p/z and cumulative production data, however, pressures representative of the true reservoir pressure are a necessity. The simple material balance method illustrated by figures 3–1 and 3–2 is virtually useless for reservoirs that have active water drives. Also, applying the method to wells in low-permeability reservoirs is difficult where pressure builds slowly after the well is shut in. Liquid production and accumulation in the wellbore add to the difficulties.

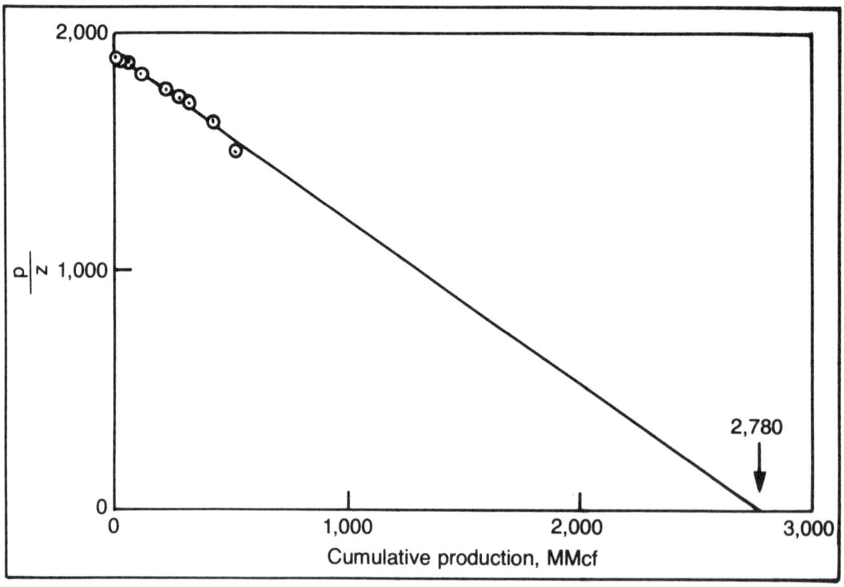

FIGURE 3–1 *Pressure-Production Curve for a Gas Well on Cartesian Coordinates*

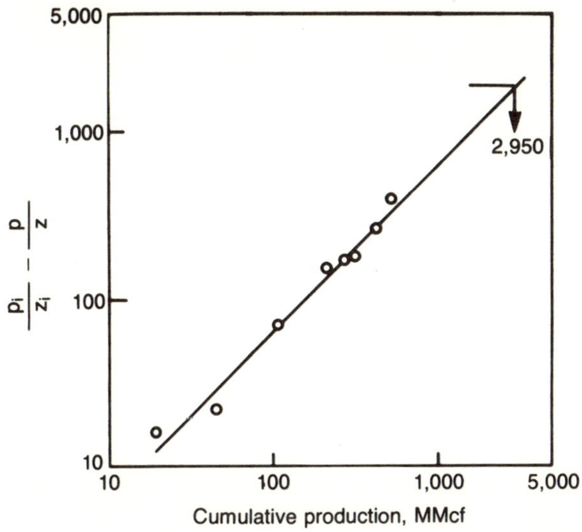

FIGURE 3–2 *Pressure-Production Curve for a Gas Well on Log-Log Coordinates*

The effect of water influx on *p/z*–cumulative-gas-production curves has been described in detail by Bruns et al. and has been reviewed from the viewpoint of obtaining the most gas from the reservoir by Agarwal et al.[1,2] The conclusions of Bruns et al. are illustrated in figure 3–3, in which the performance of a gas reservoir containing 469 billion cu ft of gas in place has been calculated for aquifer sizes 1.5 and 10 times the size of the gas reservoir. After the calculated production of 300 billion cu ft, or 64% of the gas in place, the *p/z*–cumulative-production curve for the gas reservoir in contact with the small aquifer indicates gas in place of 475 billion cu ft, or 101.3% of the gas actually in place. For a gas reservoir in contact with an aquifer 10 times the size of the gas reservoir, the curve would indicate 990 billion cu ft or 211% of the actual gas in place (see figure 3–3). Thus it was concluded that it is dangerous to extrapolate *p/z*–cumulative-production curves by using a straight line without considering the possibility of water influx.

Unfortunately, similar apparent behavior can be caused early in the production life of a well by a combination of flow rates, unsteady-state behavior, and insufficient time for buildup of the shutin pressure. This behavior is characteristic of wells in reservoirs with low permeability. At first, the well is put on production at very high rates of flow. The short-term shutin pressures show rapid decrease from the original shutin pressure. Then, the

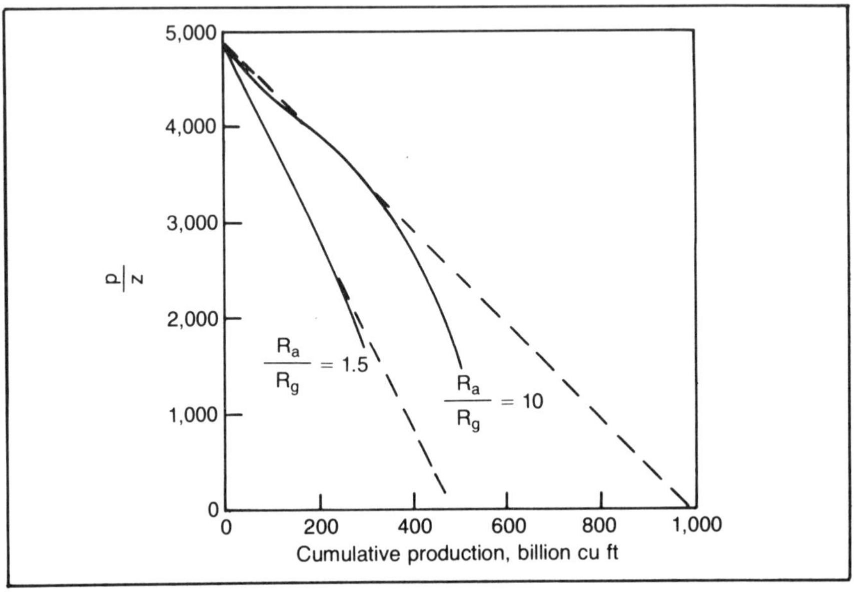

FIGURE 3–3 *Effect of Aquifier Size on* p/z*-Cumulative Production Curves*[1]

operator cuts the production rate, and shutin pressures tend to stabilize. Again, interpreting the *p/z*–cumulative-production curve becomes difficult. For an introduction to such unsteady-state behavior, the reader is referred to the treatises on pressure buildup and flow tests in wells by Matthews and Russell and by Earlougher.[3,4] Regardless of reservoir characteristics, stabilization time, or producing techniques, a minimum of 5% of the gas in place should be produced before confidence can be placed in a material balance forecast.

PROBLEMS

1. Given a pressure-production history as follows, make two types of plots and determine the initial gas in place. The specific gravity of the gas = 0.740, CO_2 = 0.00%, N_2 = 6.17%, and the reservoir temperature = 152°F.

Note: First determine z for each pressure at 152°F, then calculate p/z.

Time	*Reservoir Pressure (psia)*	*Cumulative Production (MMcf)*
0	2,066.5	0
1	1,997.1	50
2	1,782.2	479
3	1,728.4	597
4	1,647.3	722
5	1,597.6	837
6	1,533.1	945

Repeat the problem but ignore compressibility.

2. A gas reservoir at a temperature of 105°F with gas having a specific gravity of 0.610, CO_2 = 0.00%, and N_2 = 0.00% had the following production history. Calculate the original gas in place.

Reservoir Pressure (psia)	*Cumulative Production (Billion cu ft)*
2,080	0.0
1,885	6.9
1,620	14.0
1,205	23.7
888	31.0
645	36.2

REFERENCES

1. J.R. Bruns, M.J. Fetkovich, and V.C. Meitzen. "The Effect of Water Influx on p/z—Cumulative Gas Production Curves." *Jour. Pet. Tech.* (March 1965), p. 287.
2. R.G. Agarwal, R. Al-Hussainy, and H.J. Ramey Jr. "The Importance of Water Influx in Gas Reservoirs." *Trans. AIME* 234 (1965), p. 1336.
3. C.S. Matthews and D.G. Russell. *Pressure Buildup and Flow Tests in Wells.* Monograph Series, vol. 1. Dallas: Society of Petroleum Engineers of AIME (1967).
4. R.C. Earlougher Jr. *Advances in Well Test Analyses.* Mono-

graph Series, vol. 5. Dallas: Society of Petroleum Engineers of AIME (1977), p. 53.

5. H.J. Gruy and J.A. Crichton. "Critical Review of Methods Used in Estimation of Natural Gas Reserves." *Jour. of Pet. Tech.*, 11 (July 1948), p. 50.

4 Gas Measurement

In production and transportation practices, natural gas is measured by volume in cubic feet at an operating pressure and temperature and is corrected to some reference or base pressure and temperature. Also, the field engineer is involved in measurement using orifice meters and critical flow provers. The development of the principles of measurement using these devices starts with an energy balance across a general flow system as illustrated in figure 4–1 in which points 1 and 2 are any two points between which an energy balance may be considered. At point 1, the internal energy is E_1; the energy needed to move the fluid past point 1 is $p_1 V_1$; the potential energy is H_1 above the datum; and the kinetic energy is $u_1^2/2g$. Between points 1 and 2, a pump may supply energy, W, friction may cause an energy loss of W_f, and energy, q, may be gained or lost to the surroundings. The energy balance is:

$$E_1 + p_1 V_1 + H_1 + \frac{u_1^2}{2g} + q + W - W_f = E_z + p_z V_z + H_z + \frac{u_2^2}{2g} \quad (4\text{–}1)$$

If the energy balance is narrowed to 1 lb of fluid and flow through a typical, flat-plate orifice, equation 4–1 reduces to:

$$p_1 V_1 - p_2 V_2 = \frac{u_2^2}{2g} - \frac{u_1^2}{2g} \quad (4\text{–}2)$$

where:

$H_1 = H_2$ *and* $E_1 = E_2$

W_f, W, *and* $q = 0$

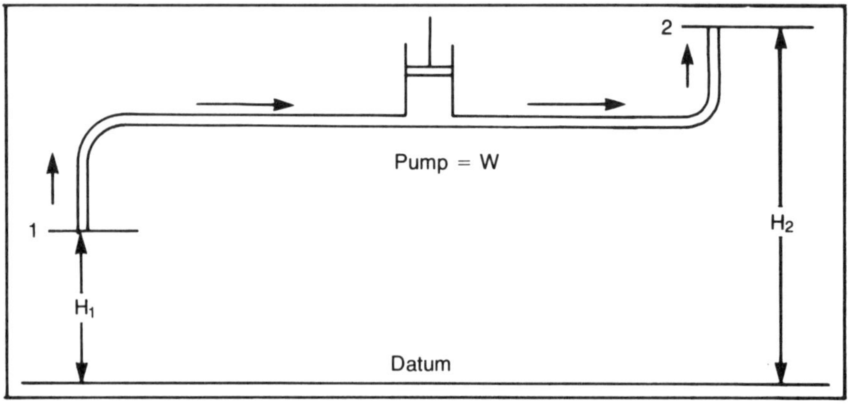

FIGURE 4–1 *General Flow System for Energy Balance*

Equation 4–2 states that, for a velocity meter, the induced change in velocity must be accompanied by a change in the static pressure. This is the principle of the orifice meter.

ORIFICE METER

If one follows the procedure of the American Gas Association Committee Report No. 3, equation 4–2 becomes:[1]

$$q_h = 218.44\, d^2 K \frac{T_b}{p_b} \sqrt{\frac{1}{T_f \gamma_g}}\ \sqrt{h_w p_f} \tag{4–3}$$

where:

q_h = flow rate, cubic ft/hr, at base conditions,
d = orifice diameter, in.,
K = discharge coefficient, which must be determined experimentally,
T_b = absolute temperature at reference or base conditions, °F + 460,
p_b = absolute pressure at reference or base conditions, psia,
T_f = absolute flowing temperature, °F + 460
γ_g = specific gravity of the flowing gas (air = 1.00)
h_w = differential pressure in inches of water
p_f = absolute static pressure, psia

For convenience, equation 4–3 is reduced to the following equations:

$$q_h = C' \sqrt{h_w p_f} \tag{4–4}$$

where:

$$C' = F_b \times F_r \times Y \times F_{pb} \times F_{tb} \times F_{tf} \times F_g \times F_{pv} \times F_m \tag{4–5}$$

where:

F_b = basic orifice factor
F_r = Reynolds number factor
Y = expansion factor
F_{pb} = pressure base factor
F_{tb} = temperature base factor
F_{tf} = flowing temperature factor
F_g = specific gravity factor
F_{pv} = supercompressibility factor
F_m = manometer factor

For actual practice in field measurement and gas transportation, this text shall adopt the tables and procedures of the Interstate Oil Compact Commission, so equations 4–4 and 4–5 become:[2]

$$q_k = F_b \times \sqrt{h_w p_f} \times F_t \times F_g \times F_{pv} \tag{4–6}$$

where:

q_k = thousands of cubic feet/24 hr (Mcfd) at a pressure base of 14.65 psia and a temperature base of 60°F
F_b = basic orifice factor
F_t = flowing temperature factor
F_g = specific gravity factor
F_{pv} = supercompressibility factor

Values for the basic orifice factor, F_b, are given in appendix B, tables B–1 and B–2, for flange- and pipe-tap types of orifice runs, respectively. Values for the flowing temperature, F_t, and the specific gravity, F_g, factors are given in tables B–4 and B–5, respectively. Values of the supercompressibility factor, F_{pv}, should be calculated using the methods given in chapter 2.

Example 4–1. Gas with a specific gravity of 0.610, a carbon dioxide and nitrogen content of 2.0 and 3.0%, respectively, is flowing through a 2-in. nominal meter run with an orifice 0.75 in. in diameter. The differential pressure is 20.5 in. of water, the flowing pressure is 624 psia, and the temperature is 105°F. Calculate the rate of flow. The meter has flange taps.

$F_b = 2.779$ (table B–1)
$F_t = 0.9594$ (table B–4)
$F_g = 1.280$ (table B–5)
$p_{pc} = 671 + 8 - 5 = 674$
$T_{pc} = 362 - 3 - 9 = 350$
$z = 0.931$
$p_{pr} = 624/674 = 0.926$
$T_{pr} = (105 + 460)/350 = 1.614$
$F_{pv} = \sqrt{1/0.931} = 1.036$
$q_k = (2.779)\sqrt{(20.5)(624)}\,(0.9594)(1.280)(1.036)$
$q_k = 400$ Mcfd at 14.65 psia and 60°F

The meter run illustrated in figure 4–2 includes upstream and downstream sections of pipe, the flanges between which the orifice is installed, the orifice plate itself, the pressure taps, and the indicating or recording instrument. The term *flange taps* is used to indicate that pressure and differential pressure measurements are made at the flanges. For *pipe taps*, the measurements are made upstream and downstream along the meter run. The use of flange taps for pressure measurement predominates since the construction and setting of the meter are easier because there is less piping and less danger of damage.

DESCRIPTION OF ORIFICE METERS

Refer to the American Gas Association Gas Measurement Committee Report No. 3 for definitive specifications and recommendations for orifice meters.[3] Figure 4–2 shows that the critical components of an orifice meter are the orifice plate, the taps for connecting instruments to measure pressure and the differential pressure across the plate, and placement of the taps with respect to the plate. Unfortunately, the figure cannot show the careful attention to construction details that allows accurate metering of natural gas.

Orifice Plates

The orifice plate should be concentric with the meter tube. It should be as flat as can be obtained commercially, the upstream edge should be square and sharp, and the surface should be as smooth as cold-finished stainless steel. The thickness should be 1/50 of the pipe diameter but no more than 1/30. Also, the thickness should be no more than 1/8 of the orifice diameter

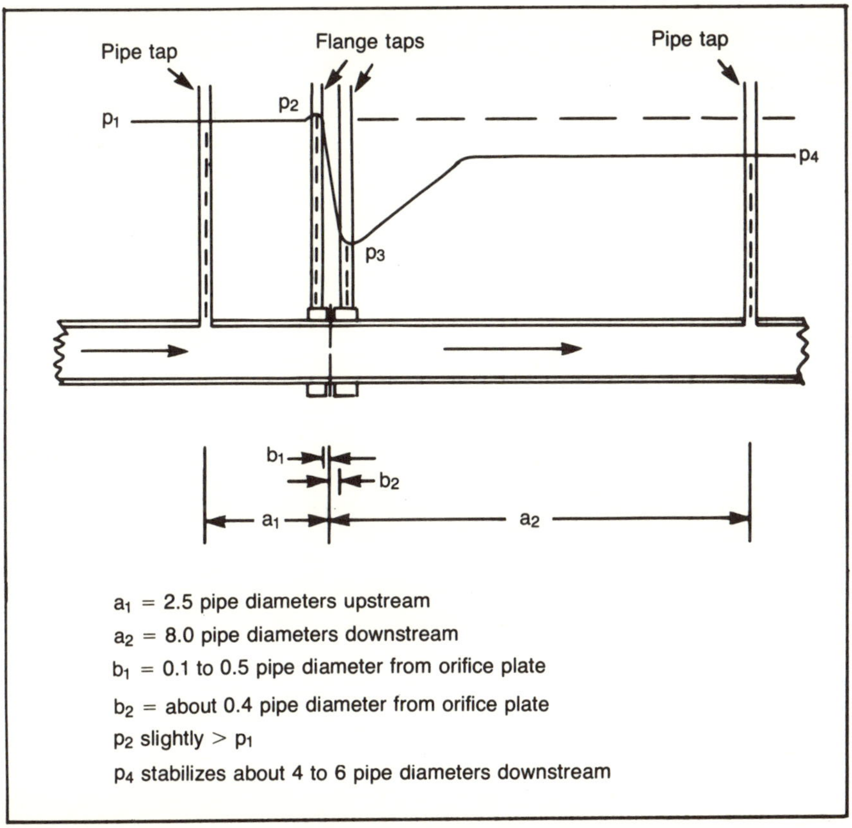

FIGURE 4–2 *Orifice Plate Showing Pressure Profile Across Plate*

and no more than ¼ of the (pipe diameter-orifice diameter)/2. The orifice plate should be kept clean at all times. If it is nicked, scarred, or out-of-shape, it should be replaced. Reference 1 presents tolerances for orifice diameters that should be followed carefully.

AGA Committee Report No. 3 recommends that the ratio of orifice diameter to pipe diameter should be between 0.15 and 0.70 for flange taps and between 0.20 and 0.67 for pipe taps.

Meter Tubes

Meter tubes upstream and downstream from the orifice plate are subject to specifications of diameter, conditions of the inside surface, and length and placement of ells upstream and downstream of the orifice. The purpose of all specifications is to create a region of steady flow conditions, both upstream and downstream from the orifice plate, and to ensure maximum metering accuracy. Needless to say, meter tubes should be free of liquids, dirt, and other foreign material.

Miscellaneous Fittings

The placement of pipe-size reducers, valves, regulators, and straightening vanes with respect to the orifice plate likewise is subject to specifications. Incidental equipment such as orifice flanges, orifice fittings, pressure tap holes, and thermometer wells should meet AGA specifications.

PULSATING FLOW

Pulsating flow consisting of sudden changes in velocity and pressure is a serious detriment to accurate gas measurement. It causes the differential pen on the chart to cover large areas of the chart with ink, or to "paint," which causes doubt as to the effective differential pressure reading. The most common sources of pulsation are:

1. Compressors
2. Pressure regulators or worn valves
3. Irregular movements of liquids in the line
4. Dumping of separators and intermitters on wells

Although it may not be possible to eliminate pulsations, it is possible to diminish the effects of pulsations on gas measurement. The meter location can be changed to the suction side of the compressor and the distance from the source can be increased, which increases the volume of gas between the source and the meter. These factors can be considered in the design and installation of a meter. The effect of pulsation on gas measurement can be minimized by using a smaller orifice so that the differential is near the upper range of the meter. If the source is a pressure regulator or valve, one may be able to adjust the regulator or replace or repair the valve so that the pulsations are diminished.

CRITICAL FLOW PROVER

Another type of gas measurement device is the critical flow prover in which measured gas is discharged under pressure through an orifice plate and into the atmosphere. The instrument wastes gas but is useful in testing gas wells that are remote from a pipeline. The testing of such wells is directed to the problem of determining if the availability of gas from the well justifies the expense of connecting the well to a market for the gas.

When gas flows through an orifice, the velocity up to a point is controlled by the differential pressure across the orifice. As the differential pressure increases—either by an increase in the upstream pressure, a decrease in the downstream pressure, or both—the velocity of the gas increases and approaches a value known as the *critical velocity*. After this critical velocity is reached, the differential pressure across the orifice no longer affects the velocity through the orifice. Critical velocity is reached when the ratio of upstream to downstream pressure is approximately 2 or more. Above the critical pressure ratio, no matter how much pressure is applied at the upstream side, the linear velocity at the *vena contracta* remains the same. This does not mean that flow through the orifice will remain the same because the density of the flowing fluid increases as the pressure increases.

For upstream pressure values greater than twice the downstream pressure ($p_1/p_2 > 2$), the rate of flow is directly proportional to the upstream pressure. Thus critical flow is said to exist at this pressure condition. The equation for calculating the rate of flow through a critical flow prover as given in the IOCC well test manual for a Bureau of Mines critical flow prover is:

$$q_k = F_p \times p_m \times F_t \times F_g \times F_{pv} \tag{4–7}$$

where:

q_k = rate of flow, Mcfd
F_p = basic orifice prover factor
p_m = pressure upstream of orifice plate, psia
F_t = flowing temperature factor
F_g = specific gravity factor
F_{pv} = supercompressibility factor

Basic orifice prover factors are given in appendix B, table B–3, and F_t and F_g are given in tables B–4 and B–5, respectively.

Example 4–2. A well is tested using a 2-in. Bureau of Mines critical flow prover. The following data are taken. Calculate the rate of flow using:

Carbon dioxide, %	0
Nitrogen, %	0
Flowing pressure, psig	270
Plate size, in.	⅜
Atmospheric pressure, psia	13.7
Flowing temperature, °F	60
Specific gravity (air = 1.00)	0.73

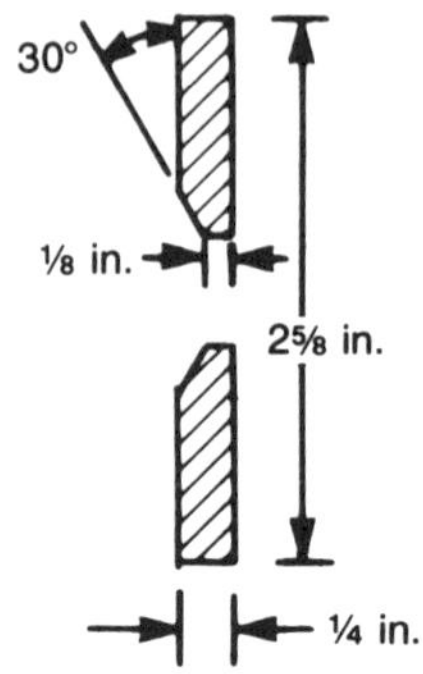

FIGURE 4–3 *Design of Orifice Plates for U.S. Bureau of Mines Type Critical Flow Prover*

Then:

$$F_p = 2.439 \text{ (table B–3)}$$
$$F_t = 1.000 \text{ (table B–4)}$$
$$F_g = 1.170 \text{ (table B–5)}$$
$$p_{pc} = 668$$
$$T_{pc} = 401$$
$$p_{pr} = 283.7/668 = 0.425$$
$$T_{pr} = (60 + 460)/401 = 1.297$$
$$z = 0.930$$
$$F_{pv} = \sqrt{1/0.930} = 1.037$$
$$q_k = \text{Mcfd} = 2.439 \times 283.7 \times 1.000 \times 1.170 \times 1.037$$
$$= 840 \text{ Mcfd at } p_b = 14.65 \text{ and } T_b = 60°\text{F}$$

Bureau of Mines Critical Flow Prover. The specifications for this 2-in. prover are given in figure 36, page 118, of Bureau of Mines Monograph 7 and the specifications for the orifice plates are shown in figure 4–3.[4] Other types of provers have been made and sold for various purposes.

PROBLEMS

1. The Bureau of Mines gives the following table of orifice prover factors and the formula for calculating rate of flow.[5] Devise a constant multiplier for determining the factors in table B–3. Note that the pressure bases for gas volumes are different.

Prover coefficients for a 2-in. critical-flow prover from the Bureau of Mines

Orifice Diameter (in.)	*Average Coefficient*	*Maximum Variation (%)*	*Orifice Diameter (in.)*	*Average Coefficient*	*Maximum Variation, (%)*
1/16	1.524	3.61	1/2	101.8	2.29
3/32	3.355	1.14	5/8	154.0	1.56
1/8	6.301	2.25	3/4	224.9	1.03
3/16	14.47	3.88	7/8	309.3	2.31
7/32	19.97	3.82	1	406.7	2.09
1/4	25.86	1.88	1 1/8	520.8	1.26
5/16	39.77	2.13	1 1/4	657.5	3.61
3/8	56.58	2.74	1 3/8	807.8	2.05
7/16	81.09	2.33	1 1/2	1,002.0	6.32

$$q' = \frac{Cp}{\sqrt{\gamma_g T}}$$

q' = Mcfd at 14.4 psia and 60°F
C = plate coefficient, given in table
p = upstream pressure, psia
γ_g = gas gravity (air = 1.00)
T = flowing temperature (°F + 460), °R

REFERENCES

1. American Gas Association. *Orifice Metering of Natural Gas.* Report No. 3. Washington, D.C.: Gas Measurement Committee, 1955.
2. Interstate Oil Compact Commission. *Manual of Back-Pressure Testing of Gas Wells.* Oklahoma City: IOCC, 1962.
3. American Gas Association, 1977.
4. E.L. Rawlins and M.A. Schellhardt. *Back-Pressure Data on Natural Gas Wells and Their Application to Production Practices.* Monograph 7. U.S. Department of Interior, Bureau of Mines, 1935.
5. Rawlins and Schellhardt, 1935.

5 Flow of Natural Gas in Circular Pipe and Annnular Conductors

The behavior of a flowing gas stream, whether in the tubing or the annular space of a well or in a pipeline is similar. The only difference between horizontal pipelines and gas wells is the pressure of the head of gas in the well, which increases the pressure drop in flowing gas wells. Flow in wells and pipelines is affected by friction factors, pipe roughness, pipe diameter, pipe length, temperature, pressure, and the properties of the gas. An accurate description of the behavior of gas flow in the form of a valid flow equation has been an engineering necessity since the early days of the gas industry. During the nearly one hundred years in which gas has been produced and transported, over two dozen flow equations have been proposed. In 1935, the Bureau of Mines published its first extensive study as Monograph 6, and a later work, Monograph 9, was published in 1956.[1,2] The equations and methods presented in these publications have been supported by subsequent developments and other laboratory experiments.

For an equation to be useful to the engineer, relatively simple determination of values for the three principal variables must be possible. These three variables—throughput flow rate, pipe diameter, and pressure drop—should be related in such a way that any one is a function of the other two and of the known properties of the gas in the pipe. They should be expressed in practical industry terminology so that they can be used easily.

DEVELOPMENT OF EQUATIONS

The unity of the principles for the flow of gas in horizontal pipelines and gas wells was demonstrated by Cullender and Smith.[3] They showed that horizontal flow in pipelines at practically constant temperature and flow in wells (either production or injection) were special cases of flow in inclined conduits.

The energy balance of equation 4–1 in differential form for flow of gas is:

$$vdp + dH + \frac{udu}{g} + dW_f = 0 \tag{5–1}$$

where:

v = specific volume
p = pressure
H = difference in elevation
u = velocity
g = conversion factor
W_f = irreversible energy loss

Equation 5–1 is usually applied to the flow of fluids, but it is equally useful in a study of static columns of gas in wells in which there is no flow. In a static gas column, the kinetic energy and irreversible-energy terms are zero. Equation 5–1, which is applicable to a unit mass of fluid, states that the energy available from the flow process is equal to the energy required to lift the unit mass of fluid through the vertical distance, dH, plus the energy involved in the change in velocity, plus the energy transformed into heat by frictional resistance. (Equation 5–1 is applicable to systems in which energy losses such as liquid slippage occur, but this text will designate dW_f as the energy transformed into heat by friction.) If the flow of gas is upward as in producing gas wells, the term dH is positive; for horizontal flow, the term dH is zero.

If equation 5–1 is applied to the upward flow of gas in a well, as illustrated in figure 5–1, and the frictional resistance term, dW_f, for a circular cross section is defined by:

$$dF = \frac{4fu^2dL}{2gD} \tag{5–2}$$

where:

f = coefficient of friction, dimensionless
D = internal diameter of pipe, ft

Equation 5–1 becomes:

$$vdp + dH + \frac{udu}{g} + \frac{4fu^2dL}{2gD} = 0 \tag{5–3}$$

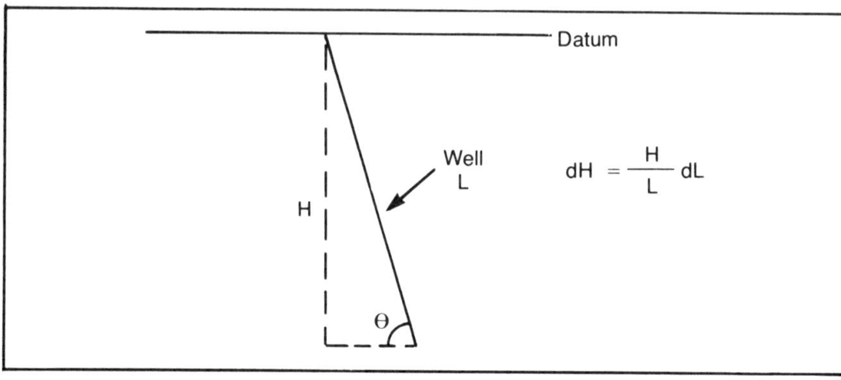

FIGURE 5–1 *Length of Pipe and Vertical Penetration for a Gas Well*[3]

Making the following substitutions:

$$u = \frac{Nv}{A}$$

$$u^2 = \frac{N^2 v^2}{A^2}$$

$$du = \frac{N}{A} dv$$

$$dH = \frac{HdL}{L}$$

$$pv = bTz$$

$$v = \frac{bTz}{p}$$

$$dv = -bTz \frac{dp}{p^2}$$

Setting the compressibility factor, z, at the pressure and temperature bases equal to 1.000, changing signs for integration from position 2 to position 1, making the proper conversion of units, and recalling from equation 2–5 that the gas constant for 1 lb of gas is:

$$b = 53.356/\gamma_g$$

equation 5–3 becomes:

$$\frac{1{,}000\ \gamma_g L}{53.356} = \int_{p_2}^{p_1} \frac{\left[p/Tz - 2.082\,\dfrac{\gamma_g {q_m}^2}{d^4 p}\right]dp}{F^2 + \dfrac{H}{L}\,\dfrac{(p/Tz)^2}{1{,}000}} \tag{5–4}$$

where:

$$F^2 = \frac{2.6665 f {q_m}^2}{d^5} = (F_r q_m)^2$$

q_m = rate of flow, MMcfd, at a standard base pressure of 14.65 psia and standard base temperature of 60°F
γ_g = specific gravity (air = 1.00)
H = difference in elevation, ft
L = length of pipe, ft
p = pressure, psia
T = absolute temperature = °F + 460
z = compressibility factor, dimensionless
d = internal diameter, in., of the circular pipe
f = coefficient of friction, dimensionless

(Equation 5–4 was reported in a rearranged form by Young but with an error in the sign for the kinetic energy term.[4])

Without making assumptions with respect to T and z, equation 5–4 does not lend itself to mathematical integration, but numerical integration over definite limits can be accomplished by the trapezoidal rule.

If:

$$\int_{p_1}^{p_n} \frac{\left[\dfrac{p}{Tz} - 2.082\,\dfrac{\gamma_g {q_m}^2}{d^4 p}\right]dp}{F^2 + \dfrac{H}{L}\,\dfrac{(p/Tz)^2}{1{,}000}} = \int_{p_1}^{p_n} I\,dp = \frac{1{,}000\ \gamma_g L}{53.356}$$

$$= \tfrac{1}{2}\,[(p_2 - p_1)(I_2 + I_1) + (p_3 - p_2)(I_3 + I_2) + \ldots (p_n - p_{n-1})(I_n + I_{n-1})] \tag{5–5}$$

Then:

$$37.484\,\gamma_g L = [(p_2 - p_1)(I_2 + I_1) + (p_3 - p_2)(I_3 + I_2) + \ldots (p_n - p_{n-1})(I_n + I_{n-1})] \quad (5\text{–}6)$$

Thus far, the only assumptions that have been made are:

1. Flow is completely turbulent.
2. The coefficient of friction, f, is a constant.
3. The compressibility of a flowing gas at the standard base pressure and temperature is 1.000.
4. Only a gas phase is flowing.

Where the variation in temperature with length, L, is known, it is necessary to select appropriate values for the length; determine p_n by trial and error so that equation 5–6 is satisfied.

Before making calculations with equations 5–6 and 5–5, one must examine the coefficient of friction, f, and show its relationship to the Reynolds number, N_{Re}. In connection with equations 5–4, 5–5, and 5–6, it was assumed that the coefficient of friction, f, was a constant. One purpose of the following section is to examine the validity and limits of this assumption.

COEFFICIENT OF FRICTION, *f*

Of the variables used in calculating flow of natural gas, the one that bothers engineers most is the coefficient of friction. In a practical sense, however, friction factors are known to the same degree of accuracy that many other variables are known. The following discussion is included to acquaint practicing engineers with the use of equations requiring a friction factor.

The coefficient of friction, f, has been defined as:

$$dF = \frac{4fu^2 dL}{2gD} \quad (5\text{–}2)$$

where:

f = coefficient of friction, dimensionless

D = internal diameter of pipe, ft

The coefficient of friction for turbulent flow in circular pipe has been the subject of an entire category of technical literature. However, space permits only a brief summary here based on the publications of Smith et al.[2,3,5]

The turbulent flow of fluids in circular conduits falls into two categories, depending upon the relationship between the boundary layer of the fluid and the roughness of the pipe wall. When the boundary layer of the fluid next to the pipe wall covers the roughness of the pipe, the fluid, in effect, is flowing in a conduit, the wall of which is composed of the fluid itself. In this case, the dominant characteristic of the container wall is the viscosity of the fluid, and the coefficient of friction is directly related to the viscosity. Since the nature of the pipe wall is masked in its effect on the flow, flow is said to be smooth even though the flow is classified as turbulent. As the velocity increases, the shear intensity at the stationary boundary layer increases, and the thickness of the boundary layer becomes less. When the roughness of the pipe wall becomes larger than the boundary layer of the fluid next to the pipe wall or the thickness of the boundary layer of fluid becomes less than the roughness of the pipe wall, the rough elements protrude into the flow stream and interfere with the flow. In this case, flow is said to be rough and the dominant characteristic of the container wall is the roughness of the wall. The coefficient of friction no longer relates to the viscosity of the fluid but is a direct function of the relative roughness of the pipe wall.

Beginning with the theoretical work of von Karman and moving through later publications, the equation for turbulent flow in smooth pipes or pipes in which the boundary layer of the fluid is thicker than the absolute roughness of the pipe is:[6–9]

$$\sqrt{1/f} = A \log N_{Re} \sqrt{f} + B_1 \tag{5–7}$$

and the equation for turbulent flow in rough pipes is:

$$\sqrt{1/f} = A \log r/k + B_2 \tag{5–8}$$

where:

A = a constant common to both equations, and B_1 and B_2 are constants
N_{Re} = Reynolds number, dimensionless
r = internal radius of pipe
k = absolute characteristic of roughness, length

Equation 5–7 for turbulent flow in smooth pipe with constants indicates that the coefficient of friction is an implicit, logarithmic function of the Reynolds number. It cannot be arranged into an explicit function of the coefficient of friction. However, the constants can be determined from flow in smooth pipes or flow in rough pipes where the wall roughness has no effect on the coefficient of friction.

Equation 5–8 has four constants, one of which is common to equation 5–7. The constant r is the internal radius of the pipe and appears with the absolute characteristic of pipe roughness, k. The principal attempt to define the roughness characteristic of equation 5–8 was made by Nikuradse, who conducted flow tests on pipes in which uniform sand grains were glued to the internal surface.[10] Nikuradse used the ratio of the pipe radius to the diameter of the sand grain as a measure of the quantity, r/k, and determined a value for B_2 in equation 5–8.

Values of the constants for equation 5–7 as determined by various investigations are given in table 5–1. Nikuradse gave a value of 3.476 for the constant B_2 in equation 5–8. Since that time, investigators have used experimental data and Nikuradse's value for B_2 to calculate an absolute roughness for commercial pipe. Taking the values of the constants from the experimental work of Smith, Miller, and Ferguson, equation 5–7 becomes:

$$\sqrt{1/f} = 4 \log N_{Re} \sqrt{f} - 0.6 \tag{5–9}$$

and equation 5–8, with the value 3.476 from Nikuradse, becomes:

$$\sqrt{1/f} = 4 \log r/k + 3.476$$

or:

$$\sqrt{1/f} = 4 \log \frac{7.4r}{k} \tag{5–10}$$

where:

r = radius of pipe
k = absolute roughness
r/k = relative roughness

TABLE 5–1.

Values of the Constants in Equation 5–7 as Determined by Various Investigators

Investigator	*Values of Constants in Equation 5–7*	
	A	B_1
von Karman[6]	4	−0.6
Nikuradse[10]	4	−0.4
Smith et al.[2]	4	−0.6

In experimental work with gas wells, Smith, Williams, and Dewees recommended an absolute roughness of 0.00065 in. for flow strings in gas wells, but later an absolute roughness of 0.0006 in. was used by Cullender and Smith for gas wells. Further work with surface pipelines resulted in an absolute roughness of 0.0007 for commercial pipe. All of these values are in essential agreement, and a value of 0.0006 in. will be used as the absolute roughness for flow in gas wells and horizontal pipe in this work.

The range of flow conditions over which equation 5–9 and equation 5–10, at an absolute roughness of 0.0006 in., are applicable is illustrated in figures 5–2 and 5–3 in which $\sqrt{1/f}$ and the coefficient of friction, f, are shown as a function of the Reynolds number (ratio of inertial forces to viscous forces) for several sizes of tubing. For example, at Reynolds numbers less than 139,000 in 1-in. tubing (ID = 1.049 in.), the coefficient of friction is determined by equation 5–9, the smooth pipe equation. At these low Reynolds numbers for natural gas, 1-in. tubing acts as if it were a smooth pipe and the natural gas is flowing in a container in which the walls consist of a boundary layer of natural gas. The boundary layer is thicker than the roughness of the pipe. At Reynolds numbers greater than 139,000, the roughness of the pipe wall becomes the dominant factor, and the coefficient of friction is independent of the Reynolds number and is determined by equation 5–10, the rough pipe equation. The lines on figures 5–2 and 5–3 for various sizes of tubing represent an absolute roughness of 0.0006 in. The minimum Reynolds number at which equation 5–10, the rough pipe equation, is applicable has been tabulated for several sizes of pipe in table 5–2.

Figures 5–2 and 5–3 indicate an abrupt change between the smooth pipe relationship (equation 5–9) in which the coefficient of friction, f, is a function of the Reynolds number and the rough pipe relationship (equation 5–10) in which the coefficient of friction is a function of the roughness of the pipe wall. If the data of Smith et al. are examined closely, a transition region is found between the two relationships that is mathematically complicated.[2,5] Previously, most investigators adopted the equation proposed by Colebrook for the transition region between smooth and rough turbulent flow.[11] However, none of the data reported by Smith et al. for the commercial pipe used in gas wells and pipelines shows a transition region similar to that reported by Colebrook. Therefore, the decision was made to drop the transition relationship from consideration in this work.

In addition to giving the minimum Reynolds number at which equation 5–10 is applicable (this is also the maximum Reynolds number at which equation 5–9 is applicable), table 5–2 gives values of $\sqrt{1/f}$ that are useful in computations with equations in which compressibility and temperature are assumed to be constant at average values. Also, the coefficient of friction and F_r values for computations by hand using equations of the type of equation 5–5 are given. F_r is defined as (see equation 5–4):

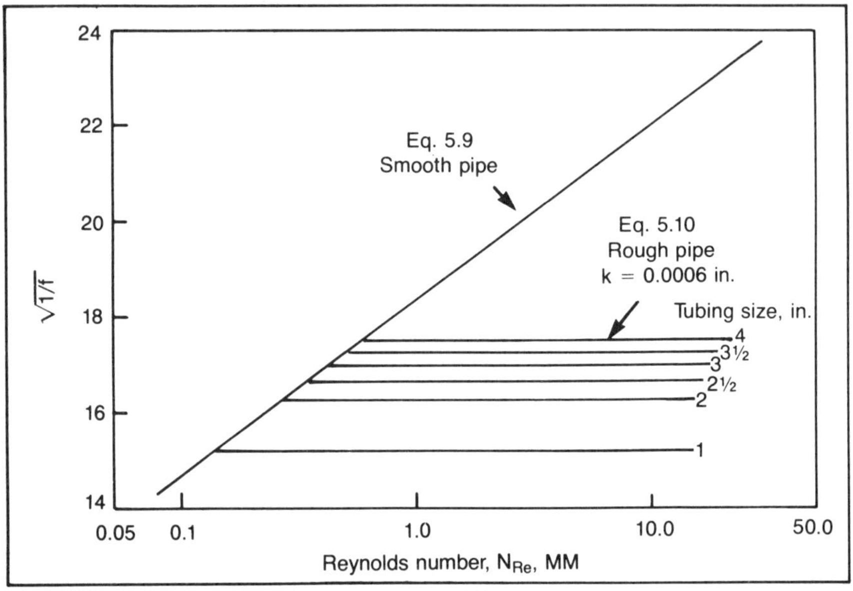

FIGURE 5–2 *Relationship Between $\sqrt{1/f}$ and Reynolds Number for Turbulent Flow*

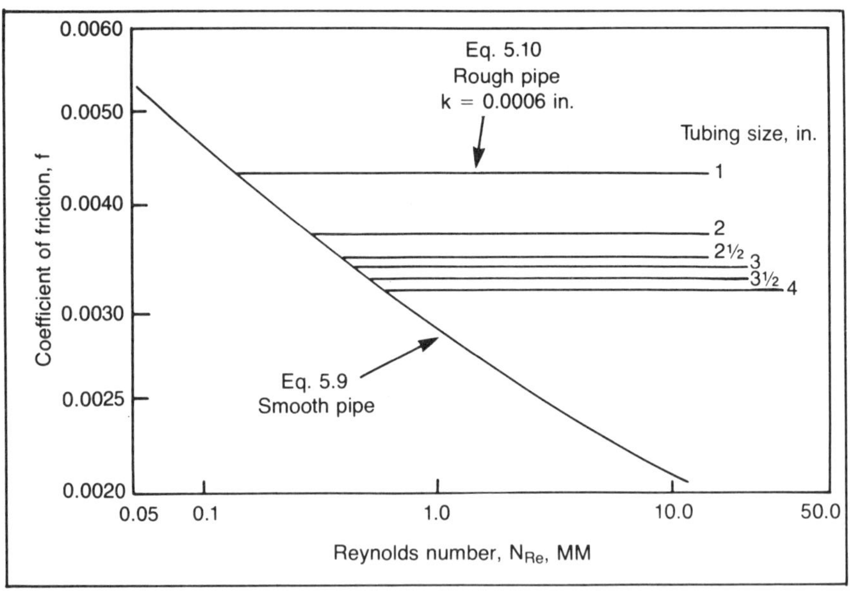

FIGURE 5–3 *Relationship Between the Coefficient of Friction and Reynolds Number for Turbulent Flow*

TABLE 5–2.
Minimum Reynolds Numbers at which Equation 5–10 is Applicable and Corresponding $\sqrt{1/f}$ and Coefficients of Friction for Various Sizes of Pipe (Absolute Roughness = 0.0006 In.)

Nominal Pipe Size (in.)	Weight (lb/ft)	OD (in.)	ID (in.)	$\sqrt{1/f}$	Coefficient of Friction (f)	F_r	Minimum Reynolds Number (N_{re})
Tubing							
1	1.80	1.315	1.049	15.24	0.004304	0.09505	139,000
1¼	2.40	1.660	1.380	15.72	0.004047	0.04643	189,000
1½	2.75	1.990	1.610	15.99	0.003912	0.03106	224,000
2	4.70	2.375	1.995	16.36	0.003736	0.01776	284,000
2½	6.50	2.875	2.441	16.71	0.003581	0.01050	355,000
3	9.30	3.500	2.992	17.06	0.003434	0.006180	445,000
3½	11.00	4.000	3.476	17.32	0.003332	0.004184	525,000
4	12.70	4.500	3.958	17.55	0.003247	0.002985	605,000
4½	16.25	4.750	4.082	17.60	0.003227	0.002755	626,000
	18.00	4.750	4.000	17.57	0.003240	0.002905	612,000
4¾	18.00	5.000	4.276	17.68	0.003198	0.002442	659,000
	21.00	5.000	4.154	17.63	0.003216	0.002633	638,000
Casing							
	13.00	5.000	4.494	17.77	0.003167	0.002146	696,000
	15.00	5.000	4.408	17.74	0.003179	0.002257	681,000
5³⁄₁₆	14.00	5.500	5.012	17.96	0.003100	0.001617	784,000
5³⁄₁₆	15.00	5.500	4.976	17.95	0.003104	0.001647	777,000
	17.00	5.500	4.892	17.92	0.003115	0.001722	763,000
	20.00	5.500	4.778	17.88	0.003129	0.001830	744,000
	23.00	5.500	4.670	17.84	0.003143	0.001942	725,000
	25.00	5.500	4.580	17.80	0.003155	0.002043	710,000
5⅝	15.00	6.000	5.524	18.13	0.003043	0.001256	872,000
	17.00	6.000	5.450	18.11	0.003050	0.001301	859,000
	20.00	6.000	5.352	18.07	0.003061	0.001363	842,000
	23.00	6.000	5.240	18.04	0.003074	0.001440	823,000
	26.00	6.000	5.140	18.00	0.003085	0.001514	806,000
6¼	20.00	6.625	6.049	18.29	0.002990	0.0009922	964,000
	22.00	6.625	5.989	18.27	0.002996	0.001018	953,000
	24.00	6.625	5.921	18.25	0.003003	0.001049	941,000
	26.00	6.625	5.855	18.23	0.003009	0.001080	930,000
	28.00	6.625	5.791	18.21	0.003015	0.001111	919,000
6¼	31.80	6.625	5.675	18.18	0.003027	0.001171	898,000
	34.00	6.625	5.595	18.15	0.003035	0.001215	885,000

Table 5–2, continued

Nominal Pipe Size (in.)	*Weight (lb/ft)*	*OD (in.)*	*ID (in.)*	$\sqrt{1/f}$	*Coefficient of Friction* (f)	F_r	*Minimum Reynolds Number* (N_{re})
6⅝	20.00	7.000	6.456	18.40	0.002954	0.0008380	1,035,000
	22.00	7.000	6.398	18.38	0.002959	0.0008578	1,025,000
	24.00	7.000	6.336	18.37	0.002964	0.0008798	1,014,000
6⅝	26.00	7.000	6.276	18.35	0.002970	0.0009018	1,003,000
	28.00	7.000	6.214	18.33	0.002975	0.0009253	992,000
	30.00	7.000	6.154	18.32	0.002981	0.0009489	982,000
	40.00	7.000	5.836	18.22	0.003011	0.001089	926,000
7¼	26.40	7.625	6.969	18.53	0.002911	0.0006872	1,125,000
7¼	29.70	7.625	6.875	18.51	0.002919	0.0007119	1,108,000
	33.70	7.625	6.765	18.48	0.002928	0.0007423	1,089,000
	38.70	7.625	6.625	18.44	0.002938	0.0007837	1,064,000
	45.00	7.625	6.445	18.40	0.002955	0.0008417	1,033,000
	26.00	8.000	7.386	18.63	0.002880	0.0005911	1,199,000
7⅝	28.00	8.125	7.485	18.66	0.002873	0.0005710	1,216,000
	32.00	8.125	7.385	18.63	0.002880	0.0005913	1,199,000
	35.50	8.125	7.285	18.61	0.002887	0.0006126	1,181,000
	39.50	8.125	7.185	18.59	0.002895	0.0006349	1,163,000
8¼	17.50	8.625	8.249	18.83	0.002822	0.0004438	1,353,000
8¼	20.00	8.625	8.191	18.81	0.002825	0.0004520	1,342,000
	24.00	8.625	8.097	18.79	0.002831	0.0004657	1,326,000
	28.00	8.625	8.003	18.77	0.002837	0.0004801	1,309,000
	32.00	8.625	7.907	18.75	0.002844	0.0004953	1,292,000
	36.00	8.625	7.825	18.73	0.002849	0.0005089	1,277,000
8¼	38.00	8.625	7.775	18.72	0.002853	0.0005174	1,268,000
	43.00	8.625	7.651	18.70	0.002861	0.0005394	1,246,000
8⅝	34.00	9.000	8.290	18.83	0.002819	0.0004382	1,360,000
	38.00	9.000	8.196	18.81	0.002825	0.0004513	1,343,000
	40.00	9.000	8.150	18.80	0.002828	0.0004579	1,335,000
8⅝	45.00	9.000	8.032	18.78	0.002836	0.0004756	1,314,000
9	36.00	9.625	8.921	18.96	0.002781	0.0003623	1,473,000
	40.00	9.625	8.835	18.95	0.002786	0.0003715	1,458,000
	43.50	9.625	8.755	18.93	0.002791	0.0003804	1,444,000
	47.00	9.625	8.681	18.91	0.002795	0.0003888	1,430,000
9	53.50	9.625	8.535	18.89	0.002804	0.0004063	1,404,000
	58.00	9.625	8.435	18.86	0.002810	0.0004189	1,386,000
9⅝	33.00	10.000	9.384	19.05	0.002756	0.0003178	1,557,000
	55.50	10.000	8.908	18.96	0.002782	0.0003637	1,471,000
	61.20	10.000	8.790	18.94	0.002789	0.0003764	1,450,000

Table 5–2, continued

Nominal Pipe Size (in.)	*Weight (lb/ft)*	*OD (in.)*	*ID (in.)*	$\sqrt{1/f}$	Coefficient of Friction (f)	F_r	Minimum Reynolds Number (N_{re})
10	32.75	10.750	10.192	19.19	0.002715	0.0002566	1,704,000
	35.75	10.750	10.136	19.18	0.002717	0.0002602	1,694,000
	40.00	10.750	10.050	19.17	0.002721	0.0002660	1,678,000
	45.50	10.750	9.950	19.15	0.002726	0.0002730	1,660,000
	48.00	10.750	9.902	19.14	0.002729	0.0002765	1,651,000
10	54.00	10.750	9.784	19.12	0.002735	0.0002852	1,630,000

$$F = (F_r q_m)^2 = \frac{2.6665 f q_m^2}{d^5}$$

or:

$$F_r = \left(\frac{2.6665 f}{d^5} \right)^{0.5} \tag{5–11}$$

AVERAGE VELOCITY AND REYNOLDS NUMBER FOR FLOW OF NATURAL GAS IN PIPE

If the average velocity in pipe is defined as:

$$\text{average velocity} = \frac{\text{volume flow rate at actual conditions}}{\text{cross-sectional area}}$$

then the average velocity, u, in feet per second is:

$$\bar{u} = 2{,}122 \frac{p_b q_m z T}{T_b p d^2}$$

or:

$$\bar{u} = 59.79 \frac{q_m z T}{p d^2} \tag{5–12}$$

The Reynolds number, a dimensionless ratio, is commonly used to characterize the conditions of fluid flow in circular pipes. Since the 1840s, a great deal of work has been performed by many investigators to determine the relationship between friction losses, pipe characteristics, fluid properties, and flow variables. Correlations for important variables were proposed long before the basic theory was formulated. In 1883, Reynolds demonstrated the existence of two distinct flow regimes. At low flow rates, he found that flow is essentially laminar. At increased flow rates, he found that random velocity components begin to occur at the center of the flow stream, gradually spreading with increasing velocity outward from the flowing core of the fluid toward the boundary. Under these circumstances, it appeared difficult to develop one single expression for the friction factor as a function of system variables. Thus the correlation of Reynolds numbers, which stated that the type of flow was a function of the magnitude of the Reynolds number, was established for the laminar flow regime, and attention was focused on the behavior in the turbulent flow regime.

In a later study, Reynolds found that the flow rate at which the transition from laminar to turbulent conditions occurred was a function of the cross-sectional area of the flowing stream, the average linear velocity, and the density and the viscosity of the fluid. For flow in pipe, the criteria of similarity was:

$$\frac{D\bar{u}\rho}{\mu}$$

This is now called the Reynolds number. The Reynolds number may be interpreted physically as the ratio of inertial forces to viscous forces. Reynolds' principle of similarity, on which the use of the number is based, states that, for flow in or around geometrically similar bodies, the flow patterns of all flows are similar if their Reynolds numbers have the same value.

It would be interesting and helpful if one could examine the laminar flow regime in detail and become familiar with the relationship between flow regime and Reynolds number. However, this text is involved solely with turbulent flow in gas wells and, in most cases, turbulent flow in natural gas pipelines.

If equation 5–11 for the average velocity is combined with equation 5–14 for the density, the Reynolds number may be calculated by the following equation:

$$N_{Re} = \frac{2.001 \times 10^4 \, q_m \gamma_g}{\mu d} \tag{5–13}$$

where the viscosity, μ, is expressed in centipoise. It is interesting to note that the pressure, temperature, and compressibility factor do not appear in equation 5–13. Therefore, the Reynolds number for flow of gas is independent of pressure, temperature, and compressibility except for the influence of pressure and temperature on the viscosity of the gas.

APPLICABILITY OF EQUATIONS FOR COEFFICIENT OF FRICTION

It has been shown from experimental work reported by many investigators that turbulent flow occurs when the Reynolds number is greater than 2,000. For Reynolds numbers greater than 2,000 to 4,000, the smooth pipe equation (equation 5–9) is applicable until the effect of the boundary layer is replaced by the effect of pipe wall roughness. This transition occurs in commercial pipe with an absolute roughness of 0.0006 in. for each size of pipe at the Reynolds numbers given in table 5–2. At Reynolds numbers larger than those in table 5–2, the rough pipe equation (equation 5–10) is applicable.

In the previous discussion, it was emphasized that the Reynolds number was independent of pressure and temperature except when viscosity was affected by pressure and temperature. It is believed that the coefficient of friction can be calculated for engineering purposes using equation 5–9, the smooth pipe equation, in the range of Reynolds numbers in which the equation is applicable.

FLOW CALCULATIONS USING EQUATIONS 5–5 AND 5–6

Producing Gas Well—Flow in a Positive Direction

Since equations have been given to describe flow and to compute the coefficient of friction, the means are available for calculating pressures in shutin and flowing gas wells, injection pressures in gas wells, and pressures along pipelines that have changes in elevation. The degree of assumptions needed to make the calculations feasible depends to some extent upon the sophistication of the means of calculation at hand. If a very large computer is available, subroutines for determining compressibility factors can be programmed internally, and solutions for equations 5–5 and 5–6 become routine. If one is limited to hand-held programmable computers, it may be necessary to estimate or calculate compressibility factors using a separate step from tables A–1, A–2, and A–3 or using equations such as equation 2–11. Therefore, the following procedures apply to the case in which compressibility factors are determined separately.

Since equations 5–5 and 5–6 were developed for flow in a producing gas well, it will be helpful to review the algebraic sign convention consistent with those conventions for producing gas wells in which the wellhead pressure is known and one desires to calculate the bottom-hole pressure. In this case, both H and L in equations 5–5 and 5–6 are considered positive in sign as indicated in figure 5–4a. Other consistent sign designations for various known pressures, injection in wells, and flow in pipelines are shown in figure 5–4.

Example 5–1; Low-pressure well producing at a high rate. This example is patterned after the example in table 30–15 in reference 12 in which a gas well ($\gamma_g = 0.615$, $H = 10{,}658$ ft, $L = 10{,}490$ ft of 2½-in. tubing, ID = 2.441 in. and 2-in. tubing, ID = 1.995 in. from 10,490 to 10,658 ft, $H = L$, surface temperature = 117°F, and temperature gradient is 5°F/1,000 ft) is assumed for the purposes of this example to be producing 11.299 MMcfd at a wellhead pressure of 500 psia and at 117°F. Calculate the flowing bottom-hole pressure.

First, the Reynolds number should be calculated using equation 5–13 at the wellhead in the 2½-in tubing:

$$N_{Re} = \frac{(2.001)(10^4)(11.299)(0.615)}{(0.0132)(2.441)} = 4.3 \times 10^6$$

The viscosity at 500 psia and 117°F was estimated to be 0.0132 centipoise. Since 4.3×10^6 is much larger than the minimum Reynolds number of 3.55×10^5 given in table 5–2, one can conclude that the friction coefficient, f, can be calculated using equation 5–10, and its value is 0.003581 according to table 5–2. The calculations using equations 5–5 and 5–6 can be made conveniently as illustrated in table 5–3. The step-by-step procedure follows.

1. Enter the well information and calculate line 1 of table 5–3. Note that p_n on line 1 is the measured wellhead pressure (in the actual example, it was an assumed value).
2. Calculate I_1 (61.358) using the definition of I given in equation 5–5 (see line 1).
3. Estimate the pressure at length L (in this case 1,000 ft) in the flow string. The first estimate of Δp can be made by calculating 37.484 $\gamma_g L$ and dividing it by $2(I_1)$ or 2×61.358. The result is 188.0, which is then used to calculate a value of 81.042 for I_2 at 1,000 ft, line 2.
4. Calculate $(\Delta p)\ (I_2 + I_1)$, which is 26,671 compared to the value that it should be of 23,053, which is 37.484 $\gamma_g L$ at 1,000 ft, line 2. Calculate a new value of $(I_2 + I_1)$ and divide it into

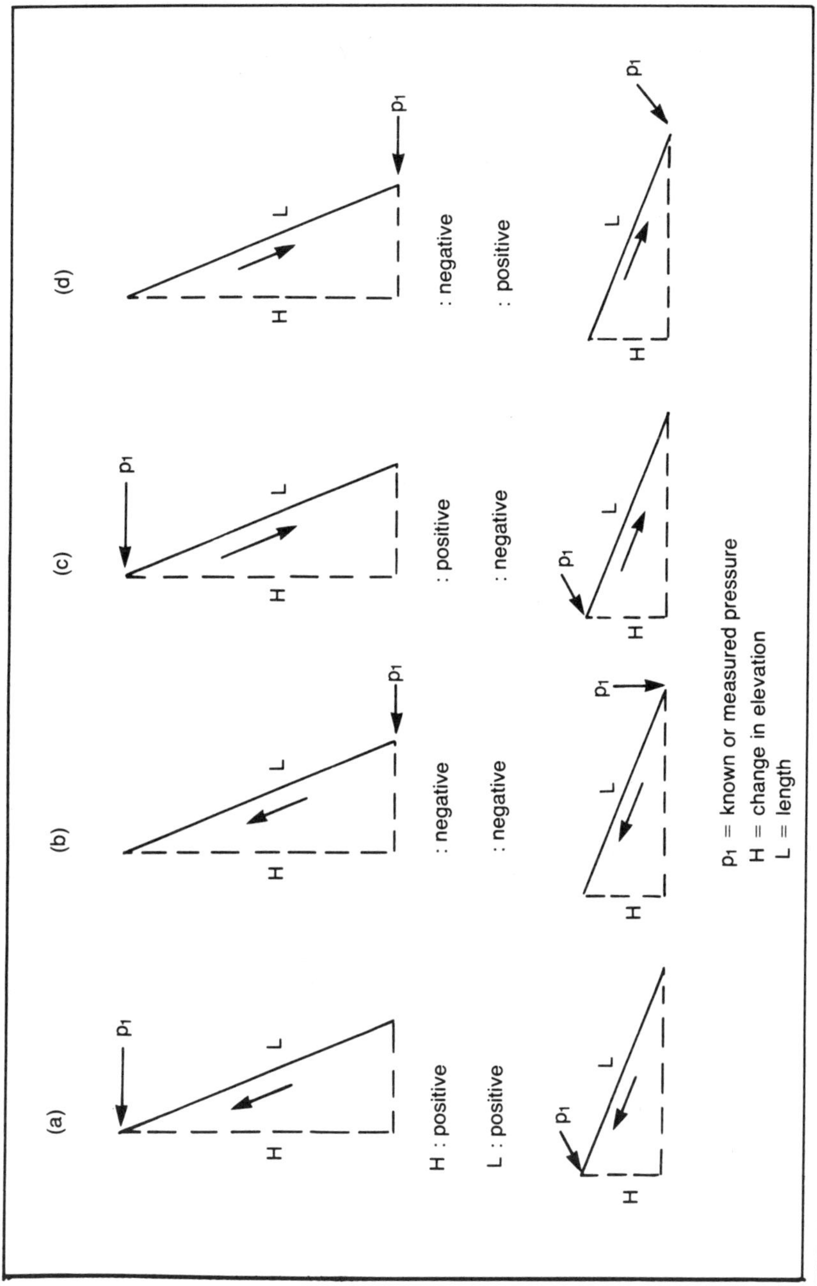

FIGURE 5–4 *Algebraic Sign Convention for Flow in Wells and Pipelines*

TABLE 5–3.

Worksheet for Calculating Subsurface Flowing Pressure Using Equations 5–5 and 5–6

Company ______ *Lease* ______ *Well No.* B *Date of test* ______

γ_g 0.615 %CO_2 2.5 %N_2 — p_{pc} 679 T_{pc} 361 *Equations used* 5–5 & 5–6

q_m 11.299 H 10,658 L 10,490 / 10,658 d 2.441 in. / 1.995 *Temperature gradient* 5°F/1,000 ft

H	L	d	p_n	t	z	I_n	Δp	$(\Delta p)x$ $(I_n + I_{n-1})$	$\Sigma(\Delta p)x$ $(I_n + I_{n-1})$	37.484x $\gamma_g L$	*Line*
0	0	2.441	500.00	117	0.937	61.358	—	—	—	0	1
1,000	1,000	2.441	688.0	122	0.924	81.042	188.0	26,671	26,671	23,053	2
1,000	1,000	2.441	664.7	122	0.927	78.608	164.7	23,052	23,052		3
2,000	2,000	2.441	800.6	127	0.916	91.083	135.9	23,061	46,113	46,105	4
3,000	3,000	2.441	920.9	132	0.908	100.469	120.3	23,044	69,156	69,158	5
4,000	4,000	2.441	1,031.6	137	0.902	107.756	110.7	23,050	92,206	92,211	6
5,000	5,000	2.441	1,135.8	142	0.898	113.463	104.2	23,051	115,256	115,263	7
6,000	6,000	2.441	1,235.4	147	0.894	118.070	99.6	23,061	138,317	138,316	8
7,000	7,000	2.441	1,331.6	152	0.893	121.630	96.2	23,059	161,376	161,369	9
8,000	8,000	2.441	1,425.2	157	0.891	124.554	93.6	23,043	184,419	184,421	10
9,000	9,000	2.441	1,516.9	162	0.891	126.819	91.7	23,051	207,470	207,474	11
10,000	10,000	2.441	1,607.2	167	0.890	128.666	90.3	23,070	230,540	230,527	12
10,490	10,490	2.441	1,650.9	169	0.890	129.426	43.7	11,279	241,819	241,822	13
10,490	10,490	1.995	1,650.9	169	0.890	60.125	—	—	241,819	241,822	14
10,658	10,658	1.995	1,683.0	170	0.889	60.856	32.1	3,883	245,702	245,695	15

23,053 for a new Δp (not shown) and continue until $\Delta p\,(I_1 + I_2)$ is in satisfactory agreement with the value of 37.484 $\gamma_g L$, line 3.

5. Go to new L of 2,000 ft and repeat the procedure.
6. In changing the flow string size as was done at L = 10,490 ft, follow the procedure shown in lines 13 and 14.

The net result of these calculations is a flowing pressure of 1,683 psia at the depth or length of 10,658 ft. In this example, the pressure at each interval is found by trial and error to be within 0.1 psia. In actual practice, this degree of precision probably is not required.

It is readily recognized that the number of calculation intervals directly affects the length of computations. If the calculations for this example were made in four steps instead of the thirteen steps illustrated in table 5–3, the resulting bottom-hole pressure would be 1,706 psia, causing a difference of 23 psi in a total pressure difference of (1,683 − 500) or 1,183 psi. The difference was caused by the nature of the change of I_n with the length, L. Examination of the values of I_n in table 5–3 shows that I_n is not a linear function of L. The closer the function is to a linear relationship, the fewer the number of steps for comparable precision. Young recommends an integration interval of 1,000 ft, as was used in example 5–1 to ensure accurate trapezoidal integration of equation 5–5.[4]

If the gas well had been drilled at an angle instead of vertically as assumed, L would have been larger than H for the nonvertical sections, and the calculations would have been carried out in the same manner.

Shutin Gas Well

In equation 5–5, the definition of I_n contained the rate of flow, q_m, in the term accounting for the change in kinetic energy, or the second term in the numerator, and in the term F in the denominator. For a shutin gas well, the rate of flow, q_m, is zero and equations 5–5 and 5–6 become:

$$\int_{p_1}^{p_n} \frac{\left(\dfrac{p}{Tz}\right) dp}{\dfrac{H}{L}\dfrac{(p/Tz)^2}{1{,}000}} = \int_{p_1}^{p_n} I dp \qquad (5\text{–}14)$$

$$= \tfrac{1}{2}\,[(p_2 - p_1)(I_2 + I_1) + (p_3 - p_2)(I_3 + I_2) + \ldots (p_n - p_{n-1})(I_n + I_{n-1})]$$

TABLE 5–4.

Worksheet for Calculating Subsurface Shutin Pressure Using Equations 5–5 and 5–6

Company ______ *Lease* ______ *Well No.* *B* *Date of test* ______

γ_g *0.615* % CO_2 *2.5* % N_2 — p_{pc} *679* T_{pc} *361* *Equations used* *5–5 & 5–6 or 5–13 & 5–14*

q_m *0* H *10,658* L *10,658* d — *Temperature gradient* *5°F/1,000 ft*

H	L	d	p_n	t	z	I_n	Δp	$(\Delta p)x$ $(I_n + I_{n-1})$	$\Sigma(\Delta p)x$ $(I_n + I_{n-1})$	$37.484x$ $\gamma_g L$	*Line*
0	0	—	1,522.0	117	0.850	322.240	—	—	—	0	1
1,000	1,000	—	1,557.8	122	0.852	315.576	35.8	22,834	22,834	23,053	2
1,000	1,000	—	1,558.0	122	0.852	318.270	36.0	23,058	23,058		3
2,000	2,000	—	1,594.4	127	0.855	314.780	36.4	23,043	46,101	46,105	4
3,000	3,000	—	1,631.2	132	0.857	311.025	36.8	23,030	69,131	69,158	5
4,000	4,000	—	1,668.5	137	0.861	308.071	37.3	23,092	92,223	92,211	6
5,000	5,000	—	1,706.1	142	0.864	304.864	37.6	23,046	115,269	115,263	7
6,000	6,000	—	1,744.1	147	0.867	301.742	38.0	23,051	138,320	138,316	8
7,000	7,000	—	1,782.5	152	0.870	298.704	38.4	23,057	161,377	161,369	9
8,000	8,000	—	1,821.2	157	0.874	296.100	38.7	23,019	184,396	184,421	10
9,000	9,000	—	1,860.4	162	0.877	293.213	39.2	23,101	207,497	207.474	11
10,000	10,000	—	1,899.9	167	0.880	290.415	39.5	23,053	230,550	230,527	12
10,658	10,658	—	1,926.1	170	0.882	288.490	26.2	15,167	245,717	245,695	13

and:

$$37.484\ \gamma_g L = [(p_2 - p_1)(I_2 + I_1) + (p_3 - p_2)(I_3 + I_2) \ldots (p_n - p_{n-1})(I_n + I_{n-1})] \tag{5–15}$$

Numerical solutions for equations 5–14 and 5–15 follow the same procedural steps given in example 5–1. It is obvious that equation 5–14 could be simplified; but with the hand-held calculator, it is more expedient to substitute $q_m = 0$ into equation 5–5, which in effect makes the computations using equation 5–14.

Example 5–2. Shutin low-pressure gas well. The gas well described in example 5–1 was shutin, and the resulting wellhead pressure was 1,522 psia. Calculate the bottom-hole pressure at a depth of 10,658 ft. Table 5–4 illustrates the calculations that were made by following steps 1 through 5 in example 5–1.

It is interesting, in comparing the calculations for a shutin gas well (table 5–4) with those for the flowing well in the previous example (table 5–3), to note that in calculations using two steps (not given in text) the bottom-hole pressure was 1,925.6 psia instead of the 1,926.1 psia calculated using twelve steps in table 5–4. In the case of the shutin gas well, the change in I_n for each 1,000-ft interval was small and of a second-order relationship. Therefore, larger calculation intervals for this shutin case do not have as much influence on the calculations as they did in example 5–1.

Injection in a Gas Well—Flow in a Negative Direction

For an injection well in which gas flows in a negative direction (downward), H is considered positive and L is considered negative (figure 5–4). Equations 5–5 and 5–6, after the sign change is made, become:

$$-\frac{1{,}000\ \gamma_g L}{53.356} = \int_{p_1}^{p_n} \frac{\left[\dfrac{p}{Tz} - 2.082\,\dfrac{\gamma_g {q_m}^2}{d^4 p}\right]}{+F^2 - \dfrac{H}{L}\dfrac{(p/Tz)^2}{1{,}000}} = \int_{p_1}^{p_n} I dp \tag{5–16}$$

and:

$$-37.484\ \gamma_g L = (p_2 - p_1)(I_2 + I_1) + (p_3 - p_2)(I_3 + I_2) + \ldots (p_n - p_{n-1})(I_n + I_{n-1}) \tag{5–17}$$

where:

$$F^2 = 2.6665 \frac{f\, q_m{}^2}{d^5}$$

The procedures for obtaining a numerical solution for equations 5–16 and 5–17 are the same as those illustrated in examples 5–1 and 5–2.

The denominator of the I term in equation 5–16 is of interest because the term, F^2, represents the energy loss caused by friction and the term on the right represents the energy gained as the gas flows down the wellbore. Since the terms have different signs, they can cancel each other exactly. This means that a discontinuity can develop for certain ranges of data. When this occurs, the pressure drop across the interval should be assumed to be zero and an inflection point occurs in the pressure traverse.

Example 5–3. Injection of gas in a well. It is proposed to inject gas into a storage well at the rate of 25.0 MMcfd at a wellhead pressure of 1,200 psi. Calculate the bottom-hole pressure at a depth of 3,000 ft. The flow string is 3-in. tubing with an ID of 2.992 in., γ_g is 0.600, CO_2 is 0%, N_2 is 2%, the wellhead temperature is 140°F, and the temperature gradient is −8°F/1,000 ft. Recalculate using the same values except the rate of injection is 10.0 MMcfd.

The Reynolds numbers for the rates of flow are 6.47×10^6 and 2.59×10^6, respectively. The results of the calculations are given in table 5–5 for both rates of flow. **Note:** for an injection rate of 25.0 MMcfd, the value of the term I_n is positive, which means that the friction loss term, F^2, in equation 5–16 is larger than the energy gained from the loss in elevation. For an injection rate of 10.0 MMcfd, I_n is negative, which means that the friction loss is less than the energy gained from the loss in elevation.

Flow in a Pipeline

The convention for the algebraic signs for elevation, H, and length, L, illustrated in figure 5–4 can be used with equations 5–5 and 5–6 to solve various gas pipeline problems using the same calculation procedures outlined previously. The advantage of doing so is that temperature gradients along the pipeline can be used without simplifying assumptions.

Example 5–4. Flow in a pipeline. Gas (γ_g = 0.600, CO_2 = 0%, and N_2 = 2.0%) is flowing at a rate of 20 MMcfd in a pipeline (ID = 10.25 in.) with an inlet pressure of 800 psia. The elevation increases by 4,000 ft over a horizontal distance of 100 mi. The temperature gradient is indicated in figure 5–5. Calculate the pressure after the gas has traveled the horizontal distance of 100 mi. The length, L, is 528,015 ft. The calculations are outlined for five-step and one-step intervals in table 5–6. In this case, both H and L are negative. The Reynolds number is 1.63×10^6 at the inlet.

TABLE 5–5.

Worksheet for Calculating Subsurface Injection Pressure Using Equations 5–5 and 5–6.

Company ______ *Lease* ______ *Well no.* ______ *Date of test* ______

γ_g *0.600* % CO_2 *0.00* % N_2 *2.00* p_{pc} *668* T_{pc} *352* *Equations used 5–5 & 5–6 or 5–16 & 5–17*

q_m *25.00* H *3,000* L *3,000* d *2.992* *Temperature gradient* $-8°F/1,000$ *ft*

H	L	d	p_n	t	z	I_n	Δp	$(\Delta p)x$ $(I_n + I_{n-1})$	$\Sigma(\Delta p)x$ $(I_n + I_{n-1})$	37.484x $\gamma_g L$	*Line*
q_m = 25.00											
0	0	2.992	1,200.0	140	0.901	116.755	—	—	—	0	1
1,000	1,000	2.992	1,103.7	132	0.903	104.849	− 96.3	−21,340	−21,340	−22,490	2
1,000	1,000	2.992	1,098.2	132	0.903	104.097	−101.8	−22,483	−22,483		3
2,000	2,000	2.992	982.6	124	0.906	90.455	−115.6	−22,490	−44,973	−44,981	4
3,000	3,000	2.992	846.8	116	0.912	75.242	−135.8	−22,502	−67,475	−67,471	5
q_m = 10.00											
0	0	2.992	1,200.0	140	0.901	−2,001.96	—	—	—	0	6
1,000	1,000	2.992	1,205.6	132	0.895	−1,674.25	+ 5.6	−20,587	−20,587	−22,490	7
1,000	1,000	2.992	1,206.1	132	0.895	−1,669.66	+ 6.1	−22,397	−22,397		8
2,000	2,000	2.992	1,213.4	124	0.888	−1,412.59	+ 7.3	−22,500	−44,897	−44,981	9
3,000	3,000	2.992	1,222.0	116	0.880	−1,209.06	+ 8.6	−22,546	−67,443	−67,471	10

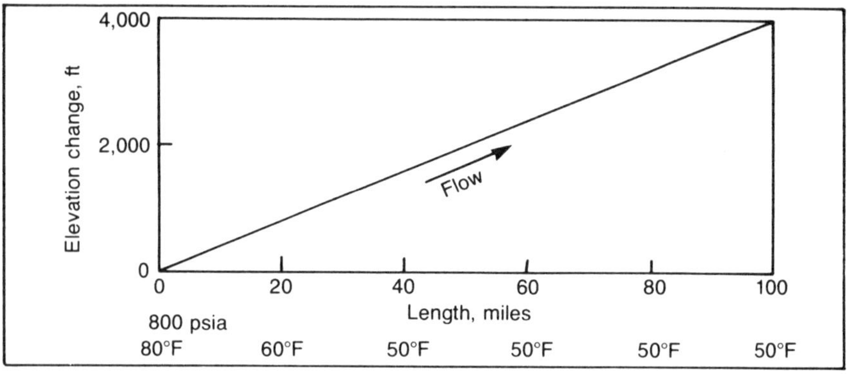

FIGURE 5–5 *Hypothetical Pipeline Showing Inlet Pressure and Temperature Profile*

For the five-step calculations (lines 1–7, table 5–6), the required pressure at 100 mi is 632.3 psia and is 631.0 psia for the one-step calculation (lines 8–10, table 5–6) for a difference of 1.3 psi.

SIZE OF INTEGRATION INTERVAL

An examination of the foregoing examples raises a question of the number of integration intervals or calculation steps needed to restrict computational errors. Young recommends an interval of 1,000 ft in gas wells, which is probably conservative.[4] Also, the integration interval depends to some extent on the sophistication of the means of calculation available to the user or, in reality, the cost of the computations. For this reason, no hard-and-fast recommendations will be made here, but the user is encouraged to set his own rules for his particular problems.

KINETIC ENERGY TERM IN FLOW EQUATIONS

From the start of efforts to calculate flow of gas in pipe, it has been customary to ignore the kinetic energy term because kinetic energy is insignificant in most flow problems. While logical, this procedure leaves an arbitrary decision that may or may not be reasonably made. Therefore, some investigation of the problem is desirable, starting with the numerator of the integrand in equation 5–5, which is:

$$\frac{p}{Tz} - 2.082\frac{\gamma_g {q_m}^2}{d^4 p}$$

TABLE 5–6.

Worksheet for Calculating Pressures Along a Pipeline Using Equations 5–5 and 5–6.

Company ______ *Line* ______ *Date of test* ______

γ_g *0.600* %CO_2 *0.00* %N_2 *2.00* p_{pc} *668* T_{pc} *352* *Equations used* *5–5 & 5–6*

q_m *20.00* H *−4000* L *−528015* d *10.25* *Temperature gradient figure 5*

H	L	d	p_n	t	z	I_n	Δp	$(\Delta p)x$ $(I_n + I_{n-1})$	$\Sigma(\Delta p)x$ $(I_n + I_{n-1})$	$37.484x$ $\gamma_g L$	*Line*
(−800)	(−105,603)	10.25	800.0	80	0.893	35,741.8	—	—	—	0	1
−800	−105,603	10.25	766.7	60	0.882	35,768.4	−33.3	−2,381,289	−2,381,289	−2,375,054	2
−800	−105,603	10.25	766.8	60	0.882	35,768.8	−33.2	−2,374,152	−2,374,152		3
−1,600	−211,206	10.25	733.5	50	0.896	35,603.0	−33.3	−2,376,680	−4,750,832	−4,750,107	4
−2,400	−316,809	10.25	700.1	50	0.885	35,419.4	−33.4	−2,372,148	−7,122,980	−7,125,161	5
−3,200	−422,412	10.25	666.4	50	0.890	35,043.3	−33.7	−2,374,593	−9,497,573	−9,500,215	6
−4,000	−528,015	10.25	632,3	50	0.895	34,541.9	−34.1	−2,372,854	−11,870,427	−11,875,269	7
(−4,000)	(−528,015)	10.25	800.0	80	0.893	35,741.8	—	—	—	0	8
−4,000	−528,015	10.25	632.3	50	0.895	34,541.9	−167.7	−11,786,576	—	−11,875,269	9
−4,000	−528,015	10.25	631,0	50	0.895	34,522.2	−169.0	−11,874,616	—	—	10

The first term represents the energy available from the expansion of the gas, and the second term represents the energy converted for changing the kinetic energy of the gas. If one makes the arbitrary decision that the kinetic energy term in equation 5–5 should be used whenever the kinetic energy term is equal to or greater than 1/100 of the available energy term, the criterion is:

$$\frac{p}{Tz} \rightleftharpoons (100)(2.082)\,\frac{\gamma_g {q_m}^2}{d^4 p}$$

or:

$$\frac{p^2 d^4}{Tz\,\gamma_g {q_m}^2} \rightleftharpoons 208.2 \tag{5–18}$$

If this criterion is applied to the preceding flow examples, the only place where the ratio of equation 5–18 is approached is from line 1 of table 5–3 where the ratio is 209.1. When the kinetic energy term is omitted, the pressure drop from a depth ($L = H$ in this example) of 1,000 ft to 0 ft is calculated at 163.6 psi instead of 164.7 psi given on line 3 of table 5–3. This is a difference of 1.1 psi for a pressure drop of 164.7 psi, or 0.7%, which is well within the tolerance of the methods. The kinetic energy term therefore could have been neglected in all of the examples.

The equations for flow in wells and pipelines given by Cullender and Smith can be derived from equations 5–5 and 5–6 by omitting the kinetic energy term and keeping the term $[(1{,}000\ \gamma_g\ L)/53.356]$ in equation 5–4 positive by dividing both sides of the equation by (−1) where necessary.[3]

FLOW AT CONSTANT TEMPERATURE AND COMPRESSIBILITY

If one assumes that the temperature and compressibility variables can be approximated using an average but constant temperature and compressibility and that the change in kinetic energy can be neglected, equation 5–5 reduces to:

$$p_{wf}^{\,2} = e^s p_{tf}^{\,2} + \frac{L}{H}(F_r\, q_m\, \bar{T}\bar{z})^2\,(e^s - 1) \tag{5–19}$$

for flowing gas wells in which pressure squared is expresed in thousands, and:

$$s = 0.037484 \frac{\gamma_g L}{\bar{T} \bar{z}}$$

For shutin gas wells, $q_m = 0$, and the equation becomes:

$$p_{ws}^2 = e^s p_{ts}^2 \tag{5–20}$$

Since $H = 0$ for horizontal pipelines, equation 5–5 becomes:

$$q_k = 1.3762 \sqrt{\frac{1}{f}} \sqrt{\frac{1}{\gamma_g}} \sqrt{\frac{1}{\bar{\bar{T}}}} \sqrt{\frac{1}{\bar{z}}} (d^{2.5}) \left(\frac{p_1^2 - p_2^2}{X} \right)^{0.5} \tag{5–21}$$

where:

X = length, miles

Solution of equation 5–19 is by trial and error on the average compressibility factor, $\bar{z}$.

WEYMOUTH FORMULA

Comparing the historic Weymouth formula with the foregoing relationships is of interest because a discussion of flow in pipe would not be complete without mentioning the relationship published by Weymouth in 1913.[13] The Weymouth formula has been used widely in engineering calculations and is still in use today. Smith, Miller, and Ferguson showed that in the Weymouth formula the coefficient of friction was related to the inside pipe diameter by:

$$f = 0.008/d^{1/3} \tag{5–22}$$

or:

$$\sqrt{1/f} = 11.18\ d^{1/6}$$

If the above are substituted into equation 5–21, and the rate of flow is changed to cubic feet per hour, one has the following:

$$q(\text{cu ft/hr}) = 641.1\, d^{8/3} \sqrt{\frac{1}{\gamma_g}} \sqrt{\frac{1}{\bar{T}}} \sqrt{\frac{1}{\bar{z}}} \left(\frac{p_1^2 - p_2^2}{X} \right)^{0.5} \tag{5-23}$$

where the length, X, is expressed in miles.

FLOW IN ANNULAR SPACES

Frequently, gas zones are completed so that production flows through the annular space between casing and tubing. Since it is impracticable to run subsurface pressure gauges in annular spaces, any knowledge of subsurface pressures, either shutin or flowing, must be gained from subsurface pressures. Therefore, it is important to have appropriate friction factors for calculating subsurface pressures in annular spaces.

If one starts with the basic differential equation (equation 5–1) for flow and develops the equations for flow in annular spaces, the only difference between the two equations is in the definition of the friction term, F_r, except for the kinetic energy term. Ignore this difference for flow in annular spaces. The friction term is:

$$F_r = \left[2.6665\, f \frac{(d_2 + d_1)}{(d_2^2 - d_1^2)^3} \right]^{0.5} \tag{5-24}$$

Take the effective diameter of a noncircular cross section as:

$$d_{eff} = (d_2 - d_1)$$

where d_2 is the inside diameter of the outer pipe and d_1 is the outside diameter of the inner pipe. Substituting the effective diameter into equation 5–10, the coefficient of friction, f, can be calculated from the following:

$$\sqrt{\frac{1}{f}} = 4 \log \frac{7.4\,(d_2 - d_1)}{2k} \tag{5-25}$$

This equation brings the value of absolute roughness, k, into question. Previously, 0.0006 in. was used for the absolute roughness of the inside of commercial pipe. Since the outside of tubing at best has upset joints and at worst collars, and since no known experimental data exist for annular spaces, it is reasonable to suggest an absolute roughness for annular spaces other than that for the inside of normal, commercial pipe. For lack of better information, use an absolute roughness of 0.001 in. for annular spaces. To this end, F_r values have been calculated for an absolute roughness of 0.001 in. for various annuli and are given in table 5–7.

SUMMARY

The development of equations for flow in gas wells and in pipelines has shown that both are special cases of the same problem. In gas wells, change in elevation is a significant variable, but in pipelines the change in elevation can be a minor variable. In either case, the pressure loss caused by friction is the principal concern of the engineer. As a result of the unity of the problems, the same calculational procedure can be used for either problem. Also, the pressure loss caused by changes in the kinetic energy of the gas can be taken into account very easily. Consequently, there is no need to neglect the kinetic energy term in any of the calculations, even when using a hand-held programmable calculator.

PROBLEMS

1. Using the appropriate equations, calculate at least five lines of data in table 5–2; start with the inside diameters of the pipe as given values.

2. Calculate the average velocity and the Reynolds number for a gas, $\gamma_g = 0.615$, $CO_2 = 4.0\%$, and $N_2 = 3.0\%$, flowing at a wellhead in 2-in. tubing at a pressure of 1,427 psia and temperature of 127°F at a rate of 2.15 MMcfd.

3. Calculate the bottom-hole flowing pressure for the well in problem 2 at a depth of 8,248 ft in 2-in. tubing. The temperature gradient is 2.5°F/1,000 ft. How many calculation steps did you use?

4. Calculate the shutin pressure for the well in problems 2 and 3 at a depth of 8,248 ft if the wellhead shutin pressure is 2,048 psia

TABLE 5–7.

F_r Values for Various Annuli (k = 0.001 in.)

Casing ID (in.)	Tubing OD (in.) 1.900	2.375	2.875	3.500	4.000	4.500	4.750	5.000
4.154	0.005082	0.006901	0.01093					
4.276	0.004576	0.006087	0.009268					
4.408	0.004107	0.005356	0.007867					
4.494	0.003838	0.004948	0.007119					
4.580	0.003593	0.004583	0.006473	0.01250				
4.670	0.003361	0.004242	0.005886	0.01086				
4.778	0.003109	0.003880	0.005281	0.009289				
4.892	0.002872	0.003544	0.004738	0.007980				
4.950	0.002761	0.003390	0.004492	0.007419				
4.976	0.002713	0.003324	0.004389	0.007187				
5.012	0.002649	0.003235	0.004251	0.006883	0.01245			
5.140	0.002438	0.002946	0.003809	0.005947	0.01012			
5.240	0.002289	0.002746	0.003509	0.005343	0.008738			
5.352	0.002137	0.002545	0.003213	0.004770	0.007506			
5.450	0.002016	0.002385	0.002983	0.004342	0.006634			
5.524	0.001931	0.002274	0.002825	0.004055	0.006074	0.01098		
5.595	0.001854	0.002175	0.002684	0.003806	0.005601	0.009783		
5.675	0.001773	0.002070	0.002538	0.003552	0.005133	0.008658		
5.791	0.001663	0.001930	0.002346	0.003226	0.004552	0.007351	0.01017	
5.836	0.001623	0.001880	0.002277	0.003111	0.004354	0.006924	0.009455	

Table 5–7, continued

Casing	*Tubing OD (in.)*							
ID (in.)	*1.900*	*2.375*	*2.875*	*3.500*	*4.000*	*4.500*	*4.750*	*5.000*
5.855	0.001607	0.001859	0.002249	0.003065	0.004274	0.006755	0.009176	
5-921	0.001551	0.001790	0.002155	0.002911	0.004012	0.006215	0.008301	
5.989	0.001497	0.001722	0.002064	0.002764	0.003768	0.005726	0.007528	
6.049	0.001452	0.001665	0.001988	0.002643	0.003570	0.005341	0.006935	0.009582
6.154	0.001376	0.001572	0.001865	0.002450	0.003260	0.004757	0.006057	0.008132
6.214	0.001336	0.001522	0.001799	0.002349	0.003100	0.004466	0.005630	0.007451
6.276	0.001296	0.001472	0.001735	0.002251	0.002947	0.004193	0.005235	0.006837
6.336	0.001259	0.001427	0.001676	0.002161	0.002810	0.003952	0.004892	0.006313
6.366	0.001241	0.001405	0.001647	0.002119	0.002745	0.003839	0.004734	0.006074
6.398	0.001222	0.001382	0.001618	0.002074	0.002678	0.003724	0.004573	0.005835
6.445	0.001195	0.001349	0.001576	0.002012	0.002584	0.003565	0.004352	0.005508
6.456	0.001189	0.001342	0.001566	0.001998	0.002563	0.003529	0.004302	0.005436
6.625	0.001099	0.001234	0.001429	0.001796	0.002266	0.003041	0.003639	0.004486
6.765	0.001032	0.001153	0.001327	0.001651	0.002057	0.002710	0.003201	0.003879
6.875	0.0009830	0.001095	0.001255	0.001549	0.001912	0.002486	0.002910	0.003486
6.969	0.0009439	0.001049	0.001198	0.001469	0.001800	0.002316	0.002692	0.003196
7.185	0.0008619	0.0009524	0.001079	0.001306	0.001577	0.001987	0.002276	0.002655
7.285	0.0008273	0.0009120	0.001030	0.001240	0.001487	0.001857	0.002116	0.002450
7.385	0.0007946	0.0008739	0.0009839	0.001178	0.001405	0.001740	0.001972	0.002268
7.386	0.0007943	0.0008736	0.0009834	0.001177	0.001404	0.001739	0.001970	0.002266

and the wellhead temperature and the temperature gradient are the same as in problems 5–2 and 5–3.

5. Assume that a natural gas, $\gamma_g = 0.700$, $CO_2 = 1.5\%$, and $N_2 = 10.0\%$, is being injected into a well in a storage reservoir at a depth of 3,500 ft. The wellhead pressure is 1,368 psia, the temperature is 125°F, the rate of injection is 28.5 MMcfd, and the flow string is a 5³⁄₁₆-in. nominal size of 25.00 lb/ft of casing. Calculate the bottom-hole flowing pressure at a depth of 3,500 ft in the flow string.

6. Use the information in the section "Kinetic Energy Term in Flow Equations" to calculate the fraction of the energy available from the expansion of the gas that is being converted to kinetic energy at the wellhead for problems 2 and 5. Could you have neglected the kinetic energy term?

7. Assume that the horizontal distance over which a pipeline with an internal diameter of 10.192 in. moves gas (characteristics given in problem 2) is 21.5 mi, the temperature is constant at 65°F, the inlet elevation is 620 ft above sea level, and the elevation is 2,650 ft after a horizontal distance of 4.5 mi from the inlet end, decreasing to 878 ft at the delivery end. The delivery pressure must be at least 314 psia. Calculate the pressure at the inlet end at several rates of flow between 1,000 and 10,000 Mcfd. Calculate the difference between the inlet pressure squared and the outlet pressure squared and plot the data on log-log paper against the rate of flow. Express the results as:

$$q_k = C(p_i^2 - p_o^2)^n$$

What are the values of C and n?

8. Change the internal diameter of the pipeline to 8.191 in. and repeat the problem. Use 100,000 Mcfd as the upper limit to the rate of flow. What is the value of n in the empirical equation?

REFERENCES

1. T.W. Johnson and W.B. Berwald. *Flow of Natural Gas Through High-Pressure Transmission Lines*. Monograph 6. U.S. Department of Interior, Bureau of Mines, 1935.

2. R.V. Smith, J.S. Miller, and J.W. Ferguson. *Flow of Natural Gas Through Experimental Pipe Lines and Transmission Lines.* Monograph 9. Department of Interior, Bureau of Mines, 1956.
3. M.H. Cullender and R.V. Smith. "Practical Solution of Gas-Flow Equations for Wells and Pipelines with Large Temperature Gradients." *Trans. AIME* 207 (1956), p. 281.
4. K.L. Young. "Effect of Assumptions Used to Calculate Bottom-Hole Pressures in Gas Wells," *Trans. AIME* 240 (1967), p. 547.
5. R.V. Smith, R.H. Williams, and E.J. Dewees. "Measurement of Resistance of Flow of Fluids in Natural Gas Wells." *Trans. AIME* 201 (1954), p. 279.
6. T. von Karman. "Turbulance and Skin Friction." *Journal Aeronaut. Science*, 1 (January 1934), p. 1.
7. H. Rouse. *Elementary Mechanics of Fluids.* New York: John Wiley & Sons Inc., 1946.
8. V.L. Streeter. "Steady Flow in Pipes and Conduits." *Engineering Hydraulics.* New York: John Wiley & Sons Inc., 1950. (Presented at 4th Hydraulics Conference, Iowa City, Iowa, June 1949).
9. B.A. Bakhmeteff. *The Mechanics of Turbulent Flow.* Princeton, N.J.: Princeton University Press, 1936.
10. J. Nikuradse. "Stromungsgesetze in Rauhen Rohren." (Laws of Fluid Flow in Rough Pipe) *V.D.I. Forschungsheft*, No. 361 (1933); *Petroleum Engineering*, vol. 11, March, May, June, July, August, 1940.
11. C.F. Colebrook. "Turbulent Flow in Pipes, with Particular Reference to the Transition Region Between the Smooth and Rough Pipe Laws." *Jour. Inst. Civil Engr.* 11, London (1938–39), p. 133.
12. R.V. Smith. "Open Flow of Gas Wells." *Petroleum Production Handbook.* Ed. T.C. Frick. New York: McGraw-Hill Book Co., 1962.
13. T.R. Weymouth. "Problems in Natural Gas Engineering." *Trans. ASME,* 34 (1913), pp. 185–231.

6 Flow of Gas in Porous Media (A Review)

This chapter reviews some of the elementary principles of flow in porous media and, as such, is by no means complete. Refer to chapter 2 of Amyx, Bass, and Whiting for a more comprehensive treatment of the subject.[1] However, an attempt has been made herein to show that approximate solutions to complicated production problems sometimes can be outlined through recourse to the elementary principles.

POROSITY

From a petroleum and natural gas production viewpoint, the two basic properties of petroleum-bearing rock are its volumetric capacity to hold fluids and its capability to transmit fluids. The quantitative measures of these properties are known as porosity and permeability, respectively. Porosity is defined as 1.0 minus the fraction of the bulk volume of the rock's solid matter. If the rock contains no fluids, porosity is the fraction of the bulk volume that is void space. Sometimes, rock contains fluids in pores that were sealed during sedimentation and, as a result, not all pore space is interconnected. Thus there is a concept of total, or absolute, porosity and effective porosity. It is common usage to use the term *porosity* for effective porosity because effective porosity represents the space that contains mobile fluids. These concepts can be expressed as:

$$\phi = \frac{V_p}{V_b} = \frac{V_b - V_{ma}}{V_b} = V_b - \frac{W}{\rho_{ma}} \qquad (6\text{–}1)$$

where:

V_p = connected pore volume

V_b = bulk volume

V_{ma} = volume of matrix materials (grain volume)

W = weight

ρ_{ma} = density of matrix materials (quartz = 2.59 to 2.66 gm/cc)

Thus, porosity may be determined by measuring any two of the three quantities—matrix volume, void volume, and bulk volume.

PERMEABILITY

The introduction to API Code RP 27 states that permeability is a property of the porous medium and is a measure of the capacity of the medium to transmit fluids.[2] The concepts of permeability or the fluid conductivity of a porous medium arose out of the work of H. Darcy in 1856. Darcy found through experimental work that the flow rate of water through sand filters was directly proportional to the area of the sand bed and to the hydraulic or pressure gradient between the inlet and outlet faces of the bed and inversely proportional to the thickness or length of the bed in the direction of flow. Darcy's relationship has been made to apply to fluids in general by including the viscosity of the fluid. The expanded version has become known as Darcy's law.

In order to make practical use of the empirical relationship known as Darcy's law, a proportionality constant known as permeability is necessary. Permeability as defined by API Code RP 27 has the units of the darcy and the millidarcy—the millidarcy being one thousandth of a darcy. A porous medium has a permeability of one darcy when a fluid of one centipoise flows at an apparent velocity of one centimeter per second under a pressure gradient of one atmosphere per centimeter under conditions of viscous flow. Specifically, the permeability constant has the dimensions of length squared (L^2) or those of an area.

Much work has been done on the permeability concept in which the limitations of Darcy's law have been defined. Numerous attempts have been made to relate permeability to the properties of a porous medium, but none has been entirely successful. Permeability measurements on thousands of core samples from oil and gas reservoir rocks have demonstrated the diversity of such media. Permeability has been found to vary widely, both laterally and vertically. The variation is generally much larger vertical to the bedding plane of stratified rocks than parallel. In spite of these variations, the concepts of permeability and Darcy's law have proved to be immensely useful in reservoir engineering.

DARCY'S LAW

The basis for a quantitative statement of permeability was formulated by H. Darcy in 1856 and was later generalized by the inclusion of viscosity, the driving forces both of pressure and of hydraulic gradients, and the application of the statement to multidimensional flow systems. Stating Darcy's law in differential form for horizontal flow is most convenient for present purposes, thus:

$$u = -\frac{k}{\mu}\frac{dp}{dl} \tag{6–2}$$

where:

k = the permeability constant
dp/dl = the pressure gradient

Note: The velocity, u, is not the actual velocity but the apparent velocity that is equivalent to q/A, or the rate of flow divided by the cross-sectional area.

If the flowing fluid is assumed to be noncompressible, equation 6–2 can be integrated as follows for linear flow:

$$\frac{q}{A}\int_{o}^{L} dl = -\frac{k}{\mu}\int_{p_1}^{p_2} dp$$

$$q = \frac{kA\,(p_1 - p_2)}{\mu L} \tag{6–3}$$

Note: If the permeability constant, k, has the units given in the quotation from API Code RP 27, the units of the remainder of the terms are fixed in equation 6–3. Thus:

$$1 \text{ darcy} = \frac{(cc/sec)(cp)(cm)}{(sq\ cm)(atm)}$$

For radial flow, the equation becomes:

$$q = \frac{2\pi\ kh\ (p_e - p_w)}{\mu \ln r_e/r_w} \tag{6–4}$$

The specimen of porous medium is assumed to be cylindrical in shape with a hole of radius, r_w, and thickness, h, as shown in figure 6–1. Again, the fluid is noncompressible and the units are those of the definition of the permeability constant.

The foregoing equations are for horizontal flow and complete saturation of a porous medium with flowing fluid. If a second fluid that is immiscible with the first fluid is introduced into the porous media, the permeability of the medium to the first fluid will change. Also, the definitions require that the flowing fluid not interact with the porous medium. As an example of interaction, a porous medium containing clays may lose all permeability when fresh water is introduced.

FLOW OF GASES IN POROUS MEDIA

When a gas is used to measure the permeability of cores in the laboratory, apparent permeability varies with the mean pressure of the measurement and the size of the gas molecules. Klinkenberg has reported variations in permeability using different gases as the flowing fluid.[3] He attributed the variations to slippage—a phenomenon observed during the flow of gases in capillaries. Consequently, slippage in porous media is related to the mean free path of the gas molecules and is controlled by temperature, pressure, and the nature of the gas. Klinkenberg and others found that the observed permeability for a gas was a straight-line function of the reciprocal of the mean pressure of the measurement, thus:

$$k_g = k_L \left(1 + \frac{b}{\overline{p}}\right) \tag{6–5}$$

where:

k_g = permeability to gas
k_L = permeability of medium to a nonreactive liquid
b = constant for a given gas in a given medium
$\overline{p}$ = mean pressure of measurement.

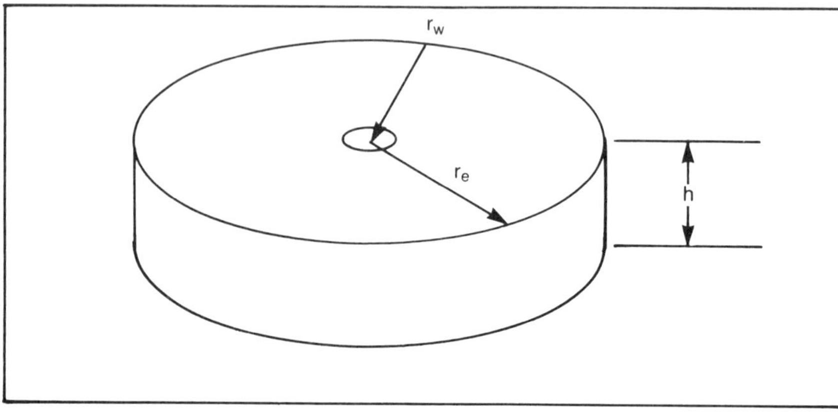

FIGURE 6–1 *Shape of Porous Medium Specimen to Illustrate Radial Flow to a Central Well Bore*

Thus, as the pressure of a measurement approaches an infinite value, the observed permeability approaches a value for a nonreactive liquid. All permeabilities are for a completely saturated medium containing a single phase.

Note: All symbols thus far have the dimensions given in the definition of permeability following equation 6–3.

EQUATIONS FOR RADIAL FLOW OF GASES

Although the relatively simple equations for steady-state flow and horizontal flow of gases in porous media have limited practical application to gas production problems, they illustrate several important problems. For this reason, the following equations are introduced.

If one starts with the definition of Darcy's law as given in equation 6–2 and derives the flow equation for radial flow of a gas, the result is:

$$q = \frac{707.8\ kh\ (p_e^2 - p_w^2)}{\bar{\mu}\bar{z}\ T_f\ \ell n\ r_e/r_w} \tag{6–6}$$

or:

$$q = 307.4 \frac{kh\ (p_e^2 - p_w^2)}{\bar{\mu}\bar{z}\ T_f\ log\ r_e/r_w} \tag{6–7}$$

where:

q = cu ft/day at 14.65 psia and 60°F
k = permeability, darcies
h = thickness, ft
p_e = pressure at external radius, r_e, psia
p_w = pressure at inner radius, r_w, psia
$\bar{\mu}$ = average viscosity, cp
$\bar{z}$ = average compressibility factor, ratio
T_f = flowing temperature, °F + 460
r = radius, ft

Equations 6–6 and 6–7 are limited to steady-state flow conditions; therefore, they do not show how pressure may vary with time. However, the equations do have some practical value in that they can be used to illustrate some of the limitations inherent in the radial flow of gas.

If one calculates the pressure in a radial direction for the flow of gas by means of equations 6–6 or 6–7 for the conditions given in table 6–1, the result is the pressure distribution given in figure 6–2, where:

k = 0.01 darcy
h = 20 ft
$\bar{\mu}$ = 0.02 cp
$\bar{z}$ = 0.90
T_f = 580°F abs
r_e = 2,640 ft
r_w = 0.25 ft
p_e = 3,000 psia
p_w = 1,000 psia
(q_k is calculated to be 11,708 Mcfd)

Pressures are shown as a function of radial distance. The most immediate reaction is that about 50% of the total pressure drop of 2,000 psi occurs within 10 ft of the wellbore. A question then presents itself—what would the rate of flow be if the wellbore were increased from the initial radius of 0.25 to 10 ft with pressures and all other conditions held constant? Calculations show that the new rate of flow would be 19,454 Mcfd. This represents a 1.66-fold increase in rate of flow for an increase in wellbore radius of 40-fold.

If one assumes that all of the conditions remain constant except the radius of the wellbore, the relationship between the rate of flow before the wellbore is enlarged (subscript 1) to that after enlargement is:

$$\frac{q_2}{q_1} = \frac{\ell n \; r_e/r_{w1}}{\ell n \; r_e/r_{w2}} = \frac{\log r_e/r_{w1}}{\log r_e/r_{w2}} \quad (6\text{–}8)$$

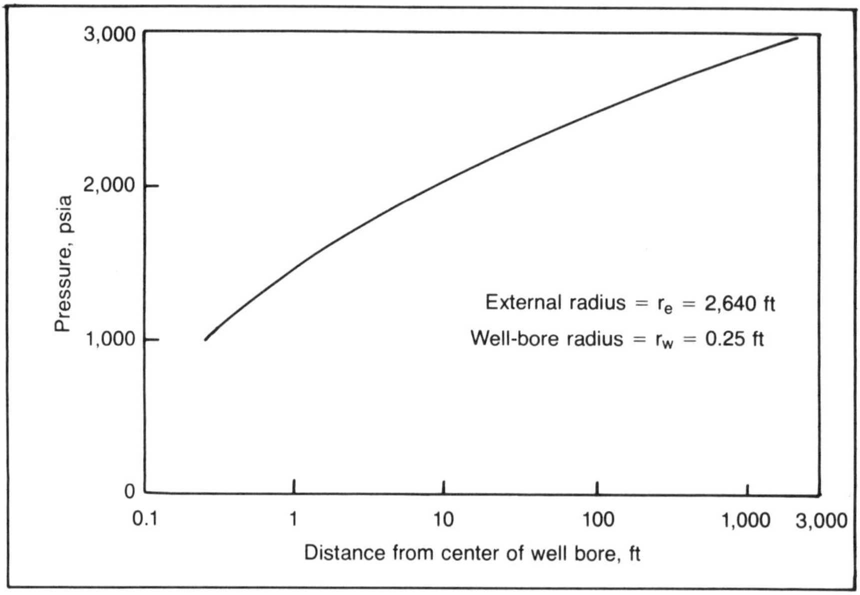

FIGURE 6–2 *Pressure Drop Along a Radius for Radial Flow of a Gas*

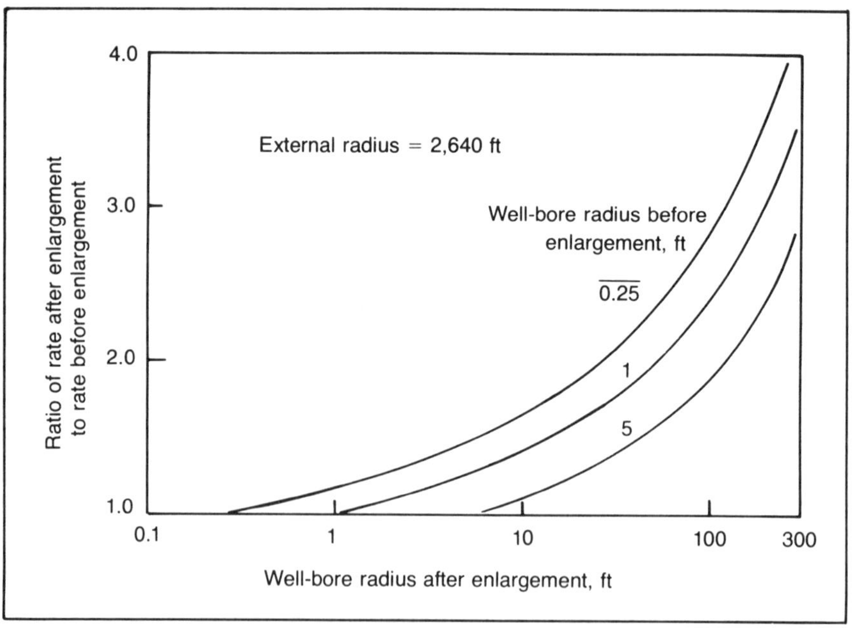

FIGURE 6–3 *Effect of Well Bore Enlargement on Producing Rate*

If all other factors remain the same, the increase in productive capacity is a function of the logarithm of the ratio of the external radius to the radius of the wellbore before and after the enlargement. This relationship of equation 6–8 is illustrated in figure 6–3. If the wellbore radius before enlargement is 0.25 ft, it is necessary to enlarge the wellbore to 26 ft to double the rate of flow provided the other conditions remain the same.

Thus, from the nature of radial flow the stimulation of wells by the enlargement of the wellbore requires relatively large increases in the size of the wellbore to bring about modest increases in the productive capacities of the well, even though most of the pressure drop occurs near the wellbore.

RADIAL FLOW OF GAS IN PARALLEL BEDS

If one considers radial flow in parallel beds as illustrated in figure 6–4, the flow through each of the separate beds is as follows:

$$q_1 = \frac{707.8\ k_1\ h_1\ (p_e^2 - p_w^2)}{\mu z\ T_f\ \ell n\ r_e/r_w}$$

$$q_2 = \frac{707.8\ k_2 h_2\ (p_e^2 - p_w^2)}{\mu z\ T_f\ \ell n\ r_e/r_w}$$

$$q_3 = \frac{707.8\ k_3 h_3\ (p_e^2 - p_w^2)}{\mu z\ T_f\ \ell n\ r_e/r_w}$$

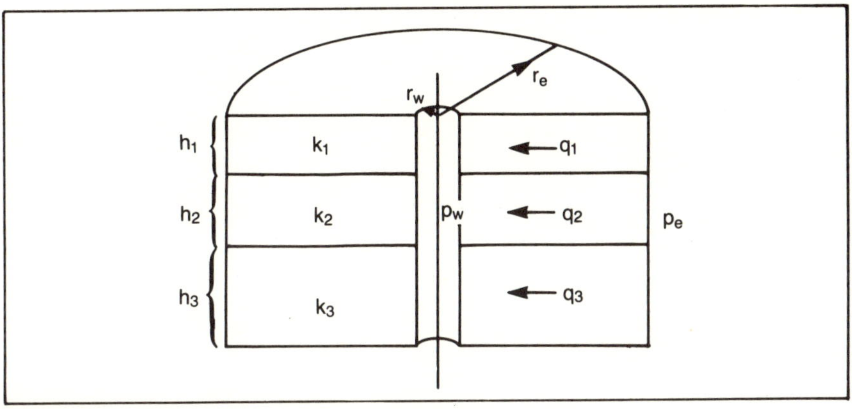

FIGURE 6–4 *Radial Flow in Parallel Beds*

Then:

$$q_1 + q_2 + q_3 = \frac{707.8\,(p_e^2 - p_w^2)}{\mu z\,\ell n\, r_e/r_w}\left[k_1h_1 + k_2h_2 + k_3h_3\right] \qquad (6\text{–}9)$$

or:

$$\bar{k} = \frac{(k_1h_1 + k_2h_2 + k_3h_3)}{(h_1 + h_2 + h_3)} \qquad (6\text{–}10)$$

where the bar over the symbol indicates an average value. In a more-general form, the result is:

$$\bar{k} = \frac{\sum_{1}^{n} k_i\, h_i}{\sum_{1}^{n} h_i} \qquad (6\text{–}11)$$

RADIAL FLOW OF GAS IN SERIES BEDS

Referring to figure 6–5 and by applying equation 6–6, the result is:

$$q = \frac{707.8\, k_2h\,(p_e^2 - p_1^2)}{\bar{\mu}_2\bar{z}_2\, T_f\, \ell n\, r_e/r_1} = \frac{707.8\, k_1h_1\,(p_1^2 - p_w^2)}{\bar{\mu}_1\bar{z}_1\, T_f\, \ell n\, r_1/r_w}$$

$$= \frac{707.8\, \bar{k}\, h\,(p_e^2 - p_w^2)}{\overline{\mu z}\; T_f\, \ell n\, r_e/r_w}$$

But:

$$(p_e^2 - p_1^2) + (p_1^2 - p_w^2) = p_e^2 - p_w^2$$

and:

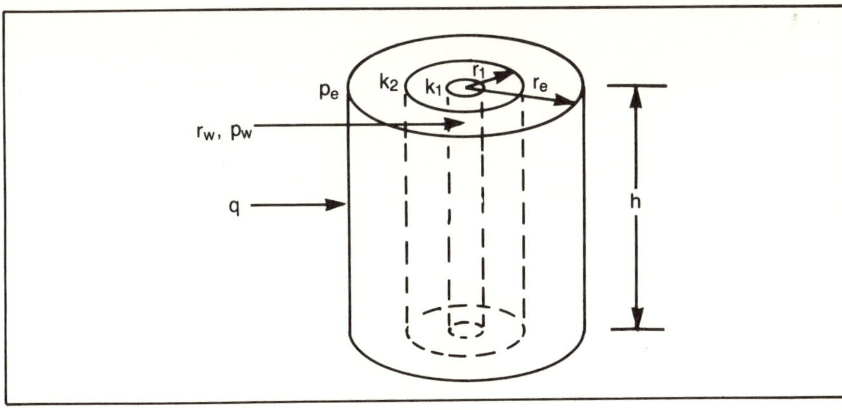

FIGURE 6–5 *Radial Flow in Series Beds*

$$\bar{k} = \frac{\bar{\mu}\bar{z}\, \ell n\, r_e/r_w}{\dfrac{\bar{\mu}_2\bar{z}_2}{k_2}\, \ell n\, r_e/r_1 + \dfrac{\bar{\mu}_1\bar{z}_1}{k_1}\, \ell n\, r_1/r_w} \tag{6–12}$$

In more general form the result is:

$$\bar{k} = \frac{\bar{\mu}\bar{z}\, \ell n\, r_e/r_w}{\sum_{1}^{n} \mu_i z_i\, \ell n\, r_i/r_{i-1}} \tag{6–13}$$

SUMMARY

The steady-state equations for flow in porous media have practical application to many gas producing problems. They are especially applicable to problems involving wellbore effects such as the effect of changes in permeability on well productivity. These applications are best illustrated by the problems that follow.

PROBLEMS

1. Calculate permeability for the following core, length = 2 cm and diameter = 1 cm. A flow of 60 cc of water/min results from an upstream pressure of 2.3 atm and a downstream pressure of 1 atm.

2. A 6-in. wellbore produces 100 Mcfd against a pressure of 500 psia. If, at an outer boundary that is 2,640-ft distant, the pressure is estimated to be 1,000 psia, and if the effective wellbore radius were increased to 125 ft, what would the resulting rate of flow be? How might the effective wellbore radius be increased to 125 ft?

3. A 6-in wellbore is drilled into a 500-md pay on a spacing of 640 acres (r_e = 2,640 ft). The drilling mud penetrated the formation to a distance of 1 ft and reduced the permeability to 10% of its original value. Assume that the gas is ideal and that the viscosity is constant. Calculate the average permeability for the well system.

REFERENCES

1. J.W. Amyx, D.M. Bass, and R.L. Whiting. *Petroleum Reservoir Engineering, Physical Properties*. New York: McGraw-Hill Book Company, 1960.
2. API Code RP 27 *Recommended Practive for Determining Permeability of Porous Media*. 3rd ed. American Petroleum Institute, Division of Production, September 1952.
3. L.J. Klinkenberg. "The Permeability of Porous Media to Liquids and Gases." *Drill. and Prod. Proc.*, API (1941), p. 200

7 Gas Well Testing

The testing of gas wells began as an effort to provide a comparative measure of the productive capacity of a well. In the early days, wells were opened to the atmosphere and allowed to "blow," and the open flow was estimated by taking impact pressure readings with a pitot tube. Although the open flows taken in this manner may have given some basis for comparing the productive capacities of wells, engineers quickly realized that the measured open flow was a poor indicator of the rate at which a gas well would produce into a pipeline under normal operating conditions. It also was a poor means for studying gas production problems. Furthermore, such open-flow testing was likely to damage the wells, was wasteful of gas, and was generally an unsafe practice.

In the 1920s, much effort was given to developing a method for measuring productive capacity by conducting tests at reasonable, controlled rates of flow. The first generally accepted test was published by Pierce and Rawlins in 1929.[1] Later, the test was set out thoroughly in Bureau of Mines Monograph 7 by Rawlins and Schellhardt in 1935.[2] The methods of Rawlins and Schellhardt quickly became standard practice and became known as the conventional back-pressure or multipoint test. Sometimes, the test is referred to as the flow-after-flow back-pressure test. Rawlins and Schellhardt found that, when the flow rates are plotted on logarithmic coordinates against the corresponding values of the difference of the squares of a formation shutin pressure, $\bar{p}_R$, and the flowing sand-face pressure, p_{wf}, the relationship may be represented emperically by a straight line. The equation is:

$$q = C\,(\bar{p}_R^{\,2} - p_{wf}^{\,2})^n \tag{7–1}$$

where C was considered a constant and sometimes is called the performance coefficient, and the exponent, n, is a constant. The industry, by common usage, refers to n as the "slope" of the back-pressure curve, even though n is the reciprocal of the mathematical slope of the line. Hereafter, n will be referred to as the exponent of the back-pressure or performance curve.

Almost from the start, so-called stabilization problems beset the early well testing procedures. At this late date, it is difficult to determine whether the stabilization problems were caused by liquid removal, by liquid accumulation in the wellbore, or by the unsteady-state behavior of the wells, but probably they were a combination of both phenomena. Testing became a procedure for cleaning the wellbore of liquids and achieving stabilization by clever testing procedures or by defining stabilization in such a way that it could be achieved easily. Typically, a flowing well was considered stabilized when the flowing wellhead pressure changed no more than a certain number of psi in a period of 15 min. This was the state of affairs until 1953–1955 when Cullender presented his paper on the isochronal method of determining the flow or performance characteristics of gas wells.[3] Essentially, Cullender pointed out that, for many wells in low-permeability reservoirs, the movement of the pressure disturbance or radius of drainage away from the wellbore upon opening the well to flow was slow. The pressure and rate effects of the movements could be observed for hours and even days after the well started flowing. Changing the rate of flow without shutting the well in only added another transient effect to the one already imposed upon the reservoir in the vicinity of the well.

Cullender proposed that the coefficient C in equation 7–1 was a variable with respect to time and that it was not a function of pressure or the rate of flow. In order to start the well on a different rate of flow, it was necessary to shut in the well and essentially erase the effect of the pressure transient on pressures in the reservoir. Thus, the isochronal back-pressure equation became:

$$q = C(t)\,(\bar{p}_R^{\,2} - p_{wf}^{\,2})^n \tag{7–2}$$

where $C(t)$ indicates that C is a function of time when each rate of flow is started with the well shut in and no prior appreciable pressure transient effects exist within the radius of drainage affected by the tests. The isochronal performance curves consist of a family of parallel curves on logarithmic coordinates in which the difference in pressures squared is plotted against the rate of flow and straight lines join the points for equal duration-of-flow time.

In effect, Cullender showed that transient pressure effects were a natural occurrence in low-permeability reservoirs and could not be negated by testing techniques. This in effect meant that the so-called absolute open flow potential of a gas well was a function of time and was affected by the pressure transients that were in existence around the wellbore at the time the well was tested. Cullender mentioned that certain gas wells stabilize so rapidly that obtaining isochronal test data was not necessary. This merely meant that the permeability in the reservoir around these wells was so high and that the pressure disturbance moved away from the well so fast that the transient effects were not observed.

At this stage of thought regarding gas well test technology, the Interstate Oil Compact Commission test manual and related test manuals were written and published.[4,5,6] Yet as these manuals were being written, there was a growing understanding of unsteady-state flow behavior as it applied both to oil and to gas wells. The first of a series of four manuals was published by the Oil and Gas Conservation Board of Alberta, Canada, in February 1964.[7] The fourth edition of *Gas Well Testing—Theory and Practice*, published in 1979, presents a comprehensive review of unsteady-state flow theory as applied to the flow of gas into gas wells.[8]

Unsteady-state flow or transient flow theory began with investigations of flow of water in aquifers and flow of hydrocarbon liquids in reservoirs. The work of Aronofsky and Jenkins, Jenkins and Aronofsky, and Bruce et al. showed that numerical solutions for transient gas flow in porous media for various rates of flow in a radial system almost coincide with the solution for transient liquid flow.[9,10,11] These findings opened the way for applying the theory of transient liquid flow to gas well behavior. A comprehensive bibliography is included in reference 8. A significant point in subsequent developments was made by Smith, who confirmed by test results that inertial turbulent flow effects could be treated as an additional rate-dependent pressure loss near the wellbore of gas wells.[12] This finding was additional to the van Everdingen skin effect.[13] Another development that cannot go unmarked was that of Al-Hussainy, Ramey, and Crawford, who introduced the concept of real gas potential $m(p)$, which includes the dimensions of pressure^2/viscosity and, according to reference 8, could be defined as a modified pressure squared.[14] In this connection, it must be pointed out that the flow of gas in porous media involves three variables—pressure, viscosity, and compressibility—when the temperature is constant. Thus, Al-Hussainy, Ramey, and Crawford have attempted to account for changes in pressure, viscosity, and compressibility simultaneously.

TYPES OF GAS WELL TESTING

The testing of gas wells falls into two general classifications depending upon the status (shutin or flowing) of the well prior to each rate of flow. If rates of flow are imposed in succession without allowing a shutin period between the flow periods, the test is a multipoint type of test. If the well is shut in between the various flow periods, the test is isochronal. A multipoint test is sometimes referred to as a three-, four-, or five-point test, depending upon the number of rates of flow used in the test. A one-point test usually means that the well was started from shutin at a predetermined rate of flow and was allowed to produce for an extended period of time. Deliverability tests are usually one-point tests run under specified conditions or tests in which the results are corrected to correspond to a specific set of conditions. More formal definitions, examples, and test procedures follow.

Multipoint Test (Flow After Flow)

Starting with a shutin well, a series of flow rates—usually in increasing sequence—are imposed on the well at fixed time intervals. The objective of the test is to determine the open-flow potential of a well or to determine the exponent (slope) to be used with a one-point test. The exponent from the original multipoint test may be used with subsequent one-point tests to determine the open-flow potential for a well as the shutin pressure decreases with production and time.

One-point Test

Starting from shutin, the well is opened to flow with producing pressures, temperatures, and rates of flow measured at specified time intervals. The duration of flow is one to three days or longer.

Isochronal Test

The isochronal test consists of a series of one-point tests, each starting with the well shut in and with the shutin pressure constant or nearly constant with time.* Producing pressures, temperatures, and rates of flow are measured and recorded at specified time intervals after the well is opened to flow. Usually the time intervals are 0.5, 1, 2, 3, 6, 24, 48, 72 hr, etc.

Information Test

The information test is a special one-point test especially designed for wells that reach stabilization slowly. The well is started from shutin to flow against a constant pressure, which can be accomplished for a small-capacity,

---

* Many investigators have called a nearly constant pressure, either shutin or flowing, a stabilized pressure, but the author prefers to use stabilization in a more-narrow sense. That is, a well is stabilized when the pressure disturbance known as the radius of investigation or influence reaches the outer boundary of the reservoir or the anticipated drainage area. This concept will be discussed later.

slow stabilization well by flowing the well at capacity into a pipeline of constant or nearly constant pressure. If a pipeline is not available, the well is opened to flow through a choke or orifice of constant size. It is important not to disturb the well by changing the choke setting or orifice size after starting the flow. Producing pressures, temperatures, and rates of flow are measured and recorded at specified time intervals after the well is opened to flow.

DEFINITIONS OF TERMS

Open-flow potential is the rate of flow that would be obtained if the bottom-hole pressure opposite the sand face were reduced to zero pressure. The open-flow potential is independent of well equipment. (The time dependency aspect should be kept in mind.)

Deliverability is the rate of flow from a well against a specified pressure (usually a working pressure at the wellhead) after a specified period of time following a specified shutin period.

Official tests are tests required by regulatory agencies for allowable purposes. The tester should obtain the test and calculation procedures from the appropriate regulatory agency before testing a well for allowable purposes. No attempt will be made to outline official testing.

PREPARATION OF A WELL FOR TESTING

The wellbore should be cleaned of liquids by flowing at a high rate to a pipeline for a period of 24 hr. (The determination of the rate of flow necessary to clean the wellbore of liquids is discussed in detail in chapter 11.) If the well does not have a pipeline connection, it may be necessary to produce the well to the atmosphere for a short period of time if such action is considered safe. Extra precautions should be taken on new wells to remove drilling mud, solids, and stimulation fluids from the wellbore. If the well has a low capacity to produce gas, extra care should be taken in swabbing the fluids from the wellbore during completion. The time and expense involved in cleaning the well can be kept to a minimum by installing a carefully sized tubing string (see chapter 11) in the well during completion. The well should be shut in for an appropriate period of time to equalize the reservoir pressure around the wellbore. Even with the largest capacity wells, the shutin period should not be less than 24 hr, if possible. For small wells, the period between completion and cleanup and connection to the pipeline should be used as the shutin period. Here, a balance must be maintained between shutin time and the economic necessity for placing a well on production.

During the shutin period, gas measurement equipment should be prepared for use. If the gas is to be measured by an orifice meter, the meter should be calibrated, the diameters and condition of the run and plate should be verified and recorded, and the differential pen should be zeroed in accordance with good meter practice. If a critical flow prover (see chapter 4) is used, it should be placed in a vertical position at the wellhead or downstream from the separator so gas will flow vertically away from the test area. If a separator is used, the rate of flow should be controlled by a production choke, and pressure should be maintained on the separator by the critical flow prover or a back-pressure regulator when an orifice meter is used. If a separator is not used, the rate of flow can be controlled at the wellhead by the critical flow prover. Always install thermometer wells at the wellhead and at gas measuring equipment so that temperatures may be measured by a thermometer or calibrated recording meter. Thermometer wells should be filled with water or lubricating oil to obtain accurate temperature measurement.

One cannot place too much emphasis on proper procedures for preparing the well for testing. The purpose and goal of the preparation is to obtain accurate test results and an absolute-open-flow potential that is indicative of the performance of the well at its best operating condition. Such tests are most valuable in studies of possible remedial work and of the installation of compressors. The natural gas engineer has the responsibility to see that highest quality tests are taken and reported.

The chief difference between testing methods for high-pressure gas wells, gas condensate wells, and low-pressure wells is the care used in taking the data and methods used in computing the results. The preparation of a high-pressure well for testing usually requires more time and patience on the part of the tester. The effect of liquids on such wells is more pronounced in high-pressure wells; consequently, special care should be used to measure gas-liquid ratios. Often, one must measure the gas-liquid ratio at each rate of flow during the multipoint or isochronal test. If the ratio is not constant, the well probably is accumulating liquid in the wellbore or unloading liquid during testing. In either case, the test probably is not acceptable and the well should be cleaned by flowing at a high rate and then retested at rates of flow high enough to keep the well free of liquid (see chapter 11).

A representative multipoint test can usually be obtained by testing the well in accordance with the following procedure:

1. Verify that the wellbore is free of liquids before shutting in the well for testing.
2. Do not start the test until the shutin pressure of the well is representative of the stablized pressure of the producing formation surrounding the wellbore.

3. The duration of each flow should be the same length of time; usually 3 hr* is sufficient.
4. The flows should be conducted in an increasing-rate sequence.

All shutin or flowing pressures should be measured using deadweight or piston gauges or other gauges of similar accuracy because bourdon gauges usually are not accurate enough for use in gas well testing. Subsurface pressures in gas wells may be measured directly by pressure gauges or calculated from surface pressures (see chapter 5). If conventional subsurface gauges are lowered into the well on a wire line, flow rates will be limited to velocities insufficient to cause the gauges to move up the producing string.

TESTING PRACTICES

The following summary lists the principal testing practices that are necessary to obtain maximum engineering and economic benefits from testing.

1. The tester should understand the purpose of testing and know what information will be recorded. He should have the proper tools for testing such as deadweight gauges, critical flow provers, and thermometers.
2. All surface pressures at the wellhead should be measured by a deadweight gauge.
3. A thermometer should be installed at the wellhead, and wellhead temperatures should be recorded at the same time that the wellhead pressure is measured.
4. The well should be conditioned for testing by removing liquids from the wellbore prior to measuring the shutin pressure (see chapter 11).
5. If an official test will be run, the procedures of the regulatory body should be followed closely, and every ethical and economic means should be used to obtain the test with the well in its best operating condition. When the results of the well influence the allowable assigned to the well, there should be no hesitation to retest if it is justified.

---

* Recently, the more-sophisticated methods for analyzing well tests have shown that wellbore storage effects in wells without packers or linear flow to stimulated wells could last easily 1 or 2 hr after the well is opened to flow or after a change is made in the rate of flow. Therefore, the author recommends a duration for each rate of flow of 3 hr.

6. Multipoint tests should be run in increasing rate of flow sequence, and each flow period should last the same period of time. During each rate of flow, wellhead flowing pressures and flow rate data should be recorded after each 15-min period to permit a determination of the degree of stabilization. In case of high-liquid-ratio wells or unusual temperature conditions, a decreasing-flow-rate sequence may be used if an increasing-flow-rate sequence results in poor alignment of points.
7. The rate of liquid production should be observed during testing at frequent intervals to ensure good liquid-gas ratio information.
8. Multipoint tests on deep, large-capacity wells in reservoirs with high temperature should be preceded by a preflow test at a high rate to ensure temperature stabilization.

SHUTIN PRESSURE

Determining the shutin pressure at the reservoir level is one of the most important measurements that can be made on a gas well. The wellhead pressure can be measured and used to calculate the pressure at the level of the reservoir, or the pressure can be measured directly by a subsurface pressure gauge. (If the gauge cannot be lowered to the reservoir level, the reservoir pressure can be determined by extrapolating a series of pressure measurements to the desired depth in the reservoir). If the shutin pressure at the reservoir level will be calculated from wellhead pressure, several precautions should be observed.

Pressure measurements at the wellhead should be taken using a deadweight tester or piston gauge at convenient time intervals until the pressure reaches a maximum or is constant (or almost constant) with time. If the well has a large production capacity and is producing froma high-pressure, high-temperature reservoir, the shutin pressure will be influenced by the temperature of the gas in the flow string. For example, a deep, high-pressure well is observed to have a wellhead pressure of 4,173 psia at a wellhead temperature of 177°F. After an extended shutin time, the wellhead pressure is 4,140 psia. The change in pressure is caused by the cooling of the column of gas in the well because no gas is being produced from the reservoir. Although the effects of temperature changes are minor and for the most part negligible for wells completed in low-pressure, low-temperature reservoirs, they become critical in initial tests upon new wells completed in deep, high-pressure, high-temperature reservoirs. The temperature behavior of such a well is given in table 7–1. The multipoint test obtained with the data in the table was

TABLE 7–1.
Temperature Behavior of a Deep Well Without Preflow to Stabilize Its Temperature

Time	*Flow Rate* (q_m)	*Wellhead Temperature (°F)*
10:20 a.m.	opened well	—
11:20	11.0	108
12:20 p.m.	13.1	110
1:20	17.7	118
2:20	23.8	128
3:20	28.4	149

unsatisfactory because of the temperature changes. Subsequently, the well was produced at a high rate for 6 hr to warm it up before the data given in table 7–2 were taken. The temperature data given in table 7–2 gave a satisfactory multipoint test.

A second source of error in calculating subsurface pressures from wellhead pressures is the presence of liquid—either hydrocarbon or water—standing in the wellbore. This problem can be avoided by taking care to flow the liquid out of the wellbore before the well is shut in or by using subsurface gauges to measure the subsurface pressures directly. Direct measurement is sometimes necessary because one cannot always free the wellbore of liquid by flowing the well before shutin.

A third problem encountered in determining reservoir pressure for a gas well is caused by unsteady-state effects in the reservoir which may persist for days, weeks or even months after the well is shut in. The pressure in the wellbore and at the wellhead of a shutin gas well may become steady in a matter of minutes or in an hour or two for large-capacity wells in high-

TABLE 7–2.
Temperature Behavior of the Same Well (Table 7–1) After Six Hours of High-Rate Flow to Stabilize Its Temperature

Time	*Flow Rate* (q_m)	*Wellhead Temperature (°F)*
2:00 p.m.	opened well	—
3:00	10.5	130
4:00	13.3	128
5:00	17.7	133
6:00	23.9	142

permeability reservoirs or may climb steadily for months after the well is shut in for a well in a very low-permeability reservoir. For the latter type of well, determining a stabilized reservoir pressure may not be an economic possibility. However, methods are described in the literature for estimating stabilized reservoir pressure from build-up pressure analyses. These methods will not be described herein, but the reader is referred to Matthews and Russell and Earlougher for the details of the methods.[15,16]

In testing gas wells, reservoir pressure is taken as the pressure at the midpoint of the productive section in the wellbore that is open to flow into the well. This point may be a large distance above or below a common datum plane in a reservoir to which reservoir pressures may be referred for other purposes.

MULTIPOINT TEST AND EXAMPLE

A multipoint test is taken so that a plot of equation 7–1:

$$q = C(\bar{p}_R^{\,2} - p_{wf}^{\,2})^n \qquad (7\text{–}1)$$

can be extrapolated for estimating the rate of flow that would be obtained if the working pressure, p_{wf}, at the sand face in the wellbore were reduced to zero. In special cases, the equation used is:

$$q = C(\bar{p}_{cs}^{\,2} - p_{cf}^{\,2})^n \qquad (7\text{–}3)$$

where flow is produced through the tubing, no packer is in the well, and pressures are measured at the wellhead on the casing. A four-point test of a constant 3-hr time duration is normally satisfactory for determining the open-flow potential of a well. Rates of flow should be taken in an increasing-rate sequence.

The four rates of flow should be evenly distributed over the test range. For average- to low-capacity gas wells, the first rate of flow should lower the pressure at the wellhead about 5% and the pressure reduction at the fourth rate of flow should be 25%. After opening the well to flow at the first rate, the test rate should be continued for 3 hr. Each succeeding rate of flow should last for the same period of time. During each flow rate, the working pressure and temperature at the wellhead, all flow rate measurements, and temperatures should be taken and recorded at the end of each 15 min. If a separator and tanks are used during testing, the rate of liquid production, both for hydrocar-

TABLE 7–3.

Field Data Sheet and Convenient Form for Multipoint Test[6]
(Reproduced by permission of the Corporation Commission of The State of Oklahoma)

TYPE TEST: 4 Point	☐ INITIAL	☐ ANNUAL	☐ SPECIAL	TEST DATE 11-3-56
COMPANY Test Example 1		CONNECTION		
FIELD	RESERVOIR		LOCATION	UNIT
COMPLETION DATE	TOTAL DEPTH	PLUG BACK TD	ELEVATION	FARM OR LEASE NAME
CSG. SIZE 7" OD	WT. 23.00# d 6.366	SET AT 8293	PERFORATIONS: FROM 8112 TO 8148	WELL NO.
TBG. SIZE 2 3/8" OD	WT. 4.70# d 1.995	SET AT 8134	PERFORATIONS: FROM 8127 TO 8132	SEC. TWP. RGE.
TYPE COMPLETION (DESCRIBE)			PACKER SET AT 8100	COUNTY
PRODUCING THRU Tubing	RESERVOIR TEMP. F 155 @ 8130	MEAN GROUND TEMP. F 60	BARO. PRESS.—Pa 14.4	STATE
L 8130	H 8130	Gg 0.625	%CO2 2.0	%N2 3.0
%H2S	PROVER METER RUN 4"	TAPS Pipe		

DATE Time of Reading	ELAP. TIME Hrs.	WELLHEAD WORKING PRESSURE Tbg. Psig	Csg. Psig	METER OR PROVER Temp. F	Pressure Psig	Diff.	Temp. F	Orifice	REMARKS (Include liquid production data: Type—API Gravity—Amount)
11-3-56									St Tk 3' 7 1/8" 60F
7:45A		1864.0	Packer						72 hour Shut-in Pressure
8:00A									Start 1st Flow
8:30A		1809.0		74	718.0	7.0	60	1.75	
9:00A		1793.0		74	720.0	8.0	59		
9:30A		1779.0		74	721.0	8.4	64		
10:00A		1772.0		74	721.0	8.3	66		St Tk 3' - 7 1/2" 60F
10:00A									Increased Rate for 2nd Flow
10:30A		1765.0		74	719.0	15.0	54	1.75	
11:00A		1742.0		74	719.0	15.5	54		
11:30A		1728.0		74	722.0	15.7	55		
12:00N		1698.0		74	729.0	17.5	56		St Tk 3' - 8 1/4" 60F
12:00N									Increased Rate for 3rd Flow
12:30P		1705.0		74	739.0	24.0	43	1.75	
1:00P		1672.0		74	740.0	25.5	46		
1:30P		1633.0		74	742.0	27.5	50		
2:00P		1608.0		74	744.0	29.3	54		St Tk 3' - 9 1/8" 60F
2:00P									Increased Rate for 4th Flow
2:30P		1630.0		74	745.0	38.0	44	1.75	
3:00P		1587.0		74	746.0	39.3	46		
3:30P		1539.0		74	749.0	41.5	52		API - 50.2° at 60°F
4:00P		1512.0		74	759.0	41.5	55		End Tk 3' - 10 1/4" 60F
									Tank Size - 12' Diameter
									1.68 Barrels/inch

PAGE 1 OF 1 DATA BY

bon and for water, should be reported. The specific gravity of the gas flowing from the prover or separator should be measured and reported. For new wells, a sample of gas from the separator or prover should be taken for analyzing hydrocarbon and nonhydrocarbon components, and for calculating heating value.

Table 7–3 is a copy of the data taken during a four-point test on a medium- to low-capacity gas well. All data recorded on table 7–3 are needed for calculating the results of the test or for understanding the results of the test. The principal information that is missing is the length of time that the well was shut in before the wellhead pressure of 1,864 psig was measured. Also, no other shutin pressures are given to indicate whether the shutin pressure was building up. From another source, we learn that the shutin time was 72 hr, but we are left to speculate the rate of pressure buildup. However, these data should have been recorded on the field data sheet.

The steps in calculating the results of a multipoint test are:

1. Calculate the rate of flow for each rate imposed on the well.
2. If the well produced hydrocarbon liquid and the specific gravity of the separator gas was measured, calculate the specific gravity of the flowing fluid using the method given in chapter 2.
3. Calculate the shutin presure, $\bar{p}_R$, the flowing bottom-hole pressures, p_{wf}, for each rate of flow, and the difference of squares between the shutin pressure and the flowing pressure for each rate of flow, $\bar{p}_R^2 - p_{wf}^2$.
4. Plot the rate of flow, q, against the corresponding pressure difference squared on log-log coordinate paper using the rate of flow as the abscissa.
5. Calculate the exponent, n, for the back-pressure curve.

The calculations are illustrated in table 7–4, in which the field test data are summarized in the upper five rows of numbers. The calculations for each of the four rates of flow are given in next four rows. The calculations for arriving at the supercompressibility factor, F_{pv}, are given in the left half of the third group of four numbers. Since the well produced hydrocarbon liquid at the ratio of 193,000 cu ft/bbl and it was assumed that all of the liquid was in the vapor phase during flow in the well, it was necessary to correct the separator gas gravity of 0.625 for the additional liquid. The gas gravity of the flowing fluid was calculated at 0.641 as indicated in the right half of the third group of four rows of numbers. (A procedure for making this calculation is given in chapter 2.) The calculations of the difference of squares for wellhead and bottom-hole pressures are given in the fourth group of numbers.

Figures 7–1 and 7–2 show the wellhead and bottom-hole performance curves, respectively, on logarithmic coordinates for the multipoint test example given in tables 7–3 and 7–4. **Note:** In comparing the curves in the two figures, the wellhead performance is steeper than the bottom-hole curve. For

TABLE 7–4.

Calculation of Results from Test Example 1 in Table 7–3[6]
(Reproduced by permission of the Corporation Commission of the State of Oklahoma)

TYPE TEST				TEST DATE	
4 Point	[X] INITIAL	[ ] ANNUAL	[ ] SPECIAL	11-3-56	
COMPANY: Test Example 1			CONNECTION		
FIELD	RESERVOIR		LOCATION		UNIT
COMPLETION DATE	TOTAL DEPTH	PLUG BACK TD	ELEVATION		FARM OR LEASE NAME
CSG. SIZE: 7" OD	WT.: 23.00#	d: 6.366	SET AT: 8293	PERFORATIONS: FROM 8112 TO 8148	WELL NO.
TBG. SIZE: 2 3/8" OD	WT.: 4.70#	d: 1.995	SET AT: 8134	PERFORATIONS: FROM 8127 TO 8132	SEC. TWP. RGE.
TYPE COMPLETION (DESCRIBE)				PACKER SET AT 8100	COUNTY
PRODUCING THRU: Tubing	RESERVOIR TEMP. F: 155 @ 8130	MEAN GROUND TEMP. F: 60		BARO. PRESS.—Pa: 14.4	STATE
L: 8130	H: 8130	γ_g: 0.625	%CO2: 2.0	%N2: 3.0	%H2S	METER RUN: 4"	TAPS: Pipe

	FLOW DATA				TUBING DATA		CASING DATA		
NO.	(LINE) X ORIFICE SIZE SIZE	PRESS. PSIG	DIFF. (INCHES)	TEMP. F	PRESS. PSIG	TEMP. F	PRESS. PSIG	TEMP. F	DURATION OF FLOW HR.
SI.					1864.0		Packer		72
1.	4 x 1.750	721.0	8.3	66	1772.0	74			2
2.	4 x 1.750	729.0	17.5	56	1698.0	74			2
3.	4 x 1.750	744.0	29.3	54	1608.0	74			2
4.	4 x 1.750	759.0	41.5	55	1512.0	74			2
5.									

RATE OF FLOW CALCULATIONS

NO.	COEFFICIENT (24—HOUR)	$\sqrt{h_w p_f}$	PRESSURE p_f	FLOW TEMP. FACTOR, F_t	GRAVITY FACTOR, F_g	SUPER COMPRESS. FACTOR, F_{pv}	RATE OF FLOW q_r
1.	17.23	78.13	735.4	0.9943	1.265	1.059	1793
2.	17.23	114.1	743.4	1.004	1.265	1.065	2659
3.	17.23	149.1	758.4	1.006	1.265	1.067	3488
4.	17.23	179.2	773.4	1.005	1.265	1.068	4192
5.							

NO.	p_{pr}	TEMP. R	T_{pr}	z
1.	1.09	526	1.49	0.891
2.	1.10	516	1.46	0.882
3.	1.13	514	1.46	0.879
4.	1.15	515	1.46	0.877
5.				

GAS LIQUID HYDROCARBON RATIO 193 q_k/BBL
API GRAVITY OF LIQUID HYDROCARBONS 50.2 DEG
SPECIFIC GRAVITY SEPARATOR GAS 0.625 | X X X X X X X X X X
SPECIFIC GRAVITY FLOWING FLUID X X X X X | 0.641
CRITICAL PRESSURE 674 PSIA | 673 PSIA
CRITICAL TEMPERATURE 353 R | 360 R

P_c 1878.4 P_c^2 3528.4

NO.	P_w	P_w^2	$P_c^2 - P_w^2$
1.		3268.5	259.9
2.		3102.6	425.8
3.		2927.4	601.0
4.		2758.1	770.3
5.			

$$(1)\ \frac{P_c^2}{P_c^2 - P_w^2} = 4.581 \qquad (2)\ \left[\frac{P_c^2}{P_c^2 - P_w^2}\right]^n = 3.136$$

$$\text{WAOF} = Q\left[\frac{P_c^2}{P_c^2 - P_w^2}\right]^n = 13\ 150$$

	$\bar{p}_{ts}$, psia	$\bar{p}_{ts}^2$ thou.	p_{tf}, psia	$(\bar{p}_{ts}^2 - p_{tf}^2)$, M	$\bar{p}_R$, psia	$\bar{p}_R^2$,M	p_{wf}, psia	$(\bar{p}_R^2 - p_{wf}^2)$, M
1	1,878.4	3,528.4	1,786.4	337.2	2,313.6	5,352.7	2,224.3	405.2
2			1,712.4	596.1			2,164.8	666.4
3			1,622.4	896.2			2,099.4	945.3
4			1,526.4	1,198.5			2,034.8	1,212.3

Wellhead curve, n = 0.647

Bottom-hole curve, n = 0.765
Open-flow potential, q_k = 12,600

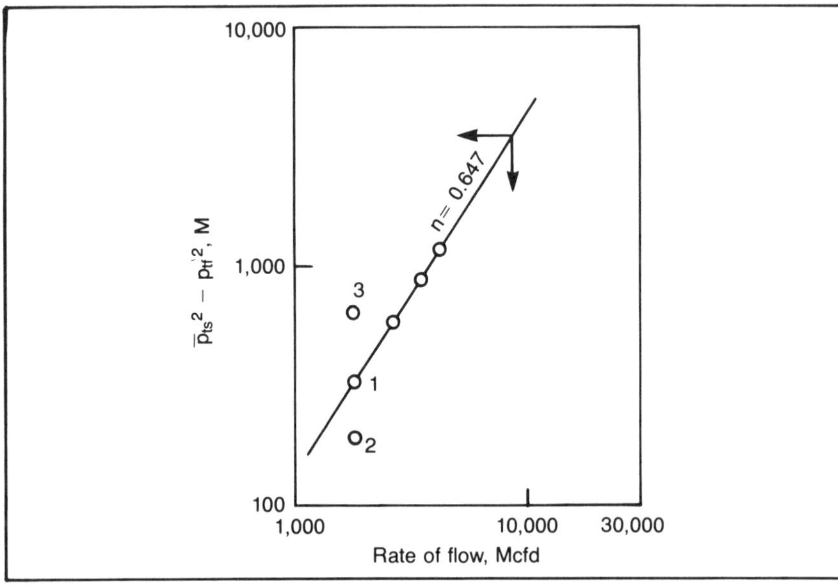

FIGURE 7–1 *Wellhead Performance Curve for Multipoint Test Example in Tables 7–3 and 7–4*

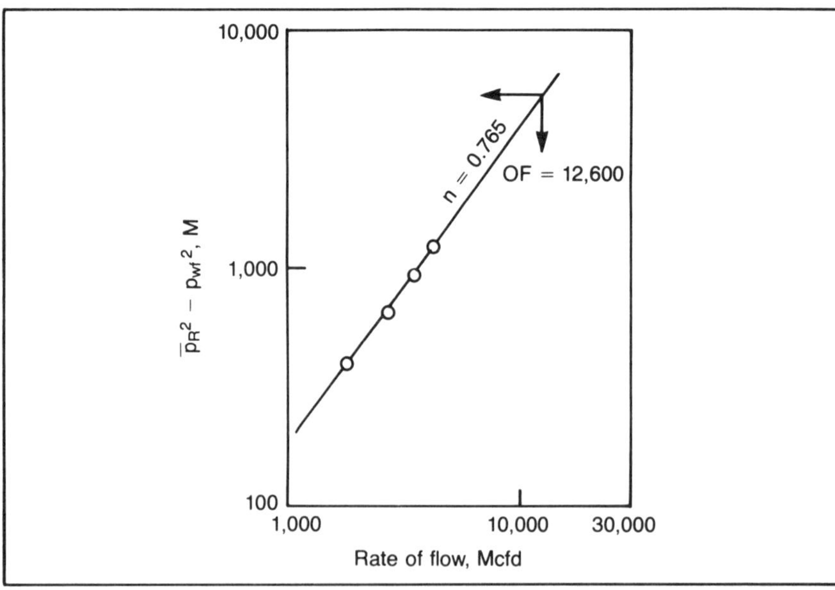

FIGURE 7–2 *Bottom-Hole Performance and Open-Flow Potential for Multipoint Test Example in Tables 7–3 and 7–4*

a given well, the values of the exponent, n, and the coefficient, C, in equation 7–1 must be determined by measuring pressures and flow rates at different rates of flow and solving for average values of n and C. If equation 7–1 is written as a logarithmic function, the result is:

$$\log q = \log C + n \log (\bar{p}_R^2 - p_{wf}^2) \qquad (7\text{–}4)$$

This forms a straight line when plotted on logarithmic coordinates. The open-flow potential is found by extrapolating the curve to $p_{wf} = 0$ or the rate of flow at $\bar{p}_R^2$ as shown in figure 7–2. Here, the average reservoir pressure, $\bar{p}_R$, was 2,313.6 psia, $\bar{p}_R^2$ was 5,352.7 (thousands), and the open-flow potential was 12,600 Mcfd.

Determination of Exponent, *n*

Since the mathematical slope of equation 7–4 is $1/n$, the value of n is given by:

$$n = \frac{\log q_2 - \log q_1}{\log (\bar{p}_R^2 - p_{wf}^2)_2 - \log (\bar{p}_R^2 - p_{wf}^2)_1}$$

$$= \frac{\log q_2/q_1}{\log (\bar{p}_R^2 - p_{wf}^2)_2/(\bar{p}_R^2 - p_{wf}^2)_1} \qquad (7\text{–}5)$$

If the determination is made graphically and the value of the difference of squares is taken exactly one cycle apart, n becomes:

$$n = \frac{\log q_2/q_1}{\log 10} = \log q_2/q_1 \qquad (7\text{–}6)$$

At $(\bar{p}_R^2 - p_{wf}^2)$ values of 2,000 and 200, the corresponding values of q are 6,000 and 1,030 Mcfd, respectively, as shown in figure 7–2, and the logarithm of 6,000/1,030 = *log* 5.825 = 0.765, as shown in figure 7–2.

Determination of Coefficient, *C*

Graphically, the value of the coefficient, C, may be determined by extrapolating the performance curve until the value of $(\bar{p}_R^2 - p_{wf}^2)$ is 1. At that point, the value of C is equal to the rate of flow. Since this would require extrapolating across two more cycles in figure 7–2, one may use the knowledge of the value of n at 0.765 and read the intersection of the performance curve at $(\bar{p}_R^2 - p_{wf}^2) = 1{,}000$ (thousands). Here, the value of q_k is 3,520 Mcfd. C is calculated from equation 7–1 as follows:

$$3{,}520 = C(1{,}000)^{0.765} \text{ and } C = 17.85$$

Or from equation 7–4:

$$\log C = \log q_k - n \log (\bar{p}_R^2 - p_{wf}^2)$$
$$\log C = \log 3{,}520 - 0.765 \log 1{,}000$$
$$C_k = \text{antilog}\,(3.547 - 2.295) = 17.85$$ (for q expressed in Mcfd and Δp^2 expressed in thousands)

Thus, the equation for the performance curve in figure 7–2 is:

$$q_k = 17.85\,(\bar{p}_R^2 - p_{wf}^2)^{0.765}$$

The difference in pressure squared is expressed in thousands and the open-flow potential is 12,700 Mcfd although it was previously estimated to be 12,600 Mcfd, reading the value from the curve. The values of n and C could be determined without the use of plots or graphs by using the method of least squares on the data given in table 7–4; however, the increase in accuracy probably would not be worth the effort.

Wellhead Performance Curve

The wellhead performance curve for the multipoint test example for tables 7–3 and 7–4 is shown in figure 7–1 where $(\bar{p}_{ts}^2 - p_{tf}^2)$ has been plotted against the corresponding rates of flow. The difference between the wellhead and bottom-hole performance curves reflects the effects of the weights of the gas columns and the friction losses on the wellhead pressure. The utility of the wellhead curve lies in the ease in calculating the data and in comparing previous tests. For example, the well in the multipoint test example should perform as well as it did in figure 7–1 over a period of time if, for each test, flow is started after a shutin period of 72 hr and the rate of flow is approximately that of the first flow rate of 1,793 Mcfd after a flow period of 2 hr. Points 2 and 3 represent hypothetical attempts to duplicate point 1. If the new performance point lies to the right of the curve as point 2 does, the performance probably has improved as a result of liquid removal from the wellbore. If the new point lies to the left of the curve as point 3 does, something has happened to the well to damage its productivity. Wellhead curves are especially useful in determining the flow rate of a well against different line pressures. The general equation for the wellhead performance curve is:

$$q = C(\bar{p}_{ts}^{\,2} - p_{tf}^{\,2})^{n} \tag{7–7}$$

The specific equation for the curve illustrated in figure 7–1 is:

$$q_k = 42.96\ (\bar{p}_{ts}^{\,2} - p_{tf}^{\,2})^{0.647}$$

The difference in pressure squared is expressed in thousands.

Critique of Multipoint Test Example

Although both the wellhead and the bottom-hole performance curves (figures 7–1 and 7–2) apply nicely to straight lines, there is no doubt that the position of the points after the first rate of flow was influenced by unsteady-state effects. Referring to table 7–3, note that, for each of the four rates of flow, the wellhead pressure decreased 7, 30, 25, and 27 psi for the respective rates of flow during the last 30 min of the flow period. The well could not have approached stabilization during each of the flow periods. Further examination of the wellhead pressures during each of the flow periods shows that the pressure decrease during the 2-hr flow periods was somewhat erratic. This could have been caused by the cyclic unloading of liquid from the well. As for the wellhead temperature, a constant temperature of 74°F was recorded throughout the testing. Experience indicates that the temperature should have increased during the testing. Thus, the multipoint test example was run under unsteady-state conditions, which means that if the tester had varied the time period from the 2-hr period used, the resultant open-flow potential would have been different from the 12,600 Mcfd reported. In other words, under the unsteady conditions that prevailed, the open-flow potential for the well in the multipoint test example was time dependent. In addition, smaller errors probably were introduced by the presence of liquids and perhaps faulty temperature measurement.

In defense of the tester, it must be pointed out that wellhead temperatures are subject to large errors under the best of conditions. Thermometer wells are seldom placed on wellheads, and even so the thermometers or other measuring devices often are influenced by ambient temperatures.

Pressure, temperature, and rate of flow data should be taken in 15-min intervals, but data was only taken every 30 min in the multipoint test example. Although this is not catastrophic in its effect, data taken every 15 min would have shed more light on the degree of stabilization and whether the well was unloading liquid from the wellbore in cycles.

SIGNIFICANCE OF OPEN-FLOW POTENTIALS

The use of open-flow potentials in connection with gas wells constitutes, for the most part, an attempt to characterize the complex behavior of wells with a single number. During the time the multipoint test was under development (1925–1935), most gas wells were relatively shallow with correspondingly low reservoir pressures and temperatures. In addition, reservoir permeabilities were relatively high and unsteady-state effects were small. Whenever such effects were recognized, the well was said not to have reached stabilization. The development of the Hugoton fields brought about an emphasis on the effects of stabilization on open-flow potentials with the result that much effort was wasted on trying to devise tests to eliminate the unsteady-state effects. The principal factors and circumstances that affect the interpretation of tests and the estimation of open-flow potentials are:

1. Unsteady-state conditions.
2. Liquids (oil, water, and condensate) in the wellbore and flow string during testing, which affect the wellhead pressures.
3. Temperature changes of the flowing or shutin column of gas.
4. Errors in calculating bottom-hole pressures.

Unfortunately, some of the factors listed above combine naturally or can be used by the tester to give inflated open-flow potentials. In many cases, the factors may cause an unwary tester to report data that result in deflated open-flow potentials. The evaluation of an open-flow potential value or test requires a judgment as to the extent that the results have been influenced by the factors listed above. While the natural gas engineer may be concerned with inflated potentials in the purchase of wells or in the evaluation of acreage from a study of offset wells, his principal responsibility is to prevent the reporting of deflated open-flow potential tests or one-point tests to state regulatory bodies. The loss in allowables and income can be substantial where allowables are restricted by poor testing. The most common causes of poor or deflated test results are inadequate cleaning of liquids from the wellbore prior to the test and failure to compare well performance with that of prior tests.

The most critical tests with respect to allowables are:

1. Multipoint tests in which production is limited to 25% of open flow.
2. Multipoint tests and one-point tests in which potential is part of the proration formula.

3. Deliverability tests in fields in which deliverability is part of the proration formula.
4. The one-point test in which the allowable is limited to the test rate reported on the official test.
5. Shutin pressure tests in which pressure is part of the proration formula or is used in determining a contractural quantity of gas to be purchased.

Open-flow potentials or, more accurately, the concept of a maximum producing capacity for a gas well can be valuable when considering the effect of delivery pressure on the capacity of a well to deliver gas. Starting with equation 7–7, a relationship can be developed between wellhead flowing pressure as a fraction of shutin wellhead pressure and rate of flow as a fraction of the wellhead open-flow potential.

$$q = C(\bar{p}_{ts}^2 - p_{tf}^2)^n \tag{7–7}$$

The open-flow potential or maximum producing capacity at the wellhead is:

$$q_{of} = C(\bar{p}_{ts}^2)^n \tag{7–8}$$

where q_{of} is the open-flow potential at the wellhead or maximum producing capacity when the wellhead pressure is zero. Then:

$$\frac{q}{q_{of}} = \frac{C(\bar{p}_{ts}^2 - p_{tf}^2)^n}{C(\bar{p}_{ts}^2)^n}$$

or:

$$\frac{q}{q_{of}} = \left[1 - \left(\frac{p_{tf}}{\bar{p}_{ts}}\right)^2\right]^n \tag{7–9}$$

A plot of equation 7–9 is shown in figure 7–3 in which the flowing pressure is a fraction of the shutin pressure, and the rate of flow is a fraction of the maximum producing capacity or open-flow potential.

Figure 7–3 is of very practical value when working with expected rates of flow against various line pressures for a given well. For example, assume a well has been producing, without restriction, about 150 Mcfd against a pipeline pressure of 1,000 psia for a long period of time. The well has a current

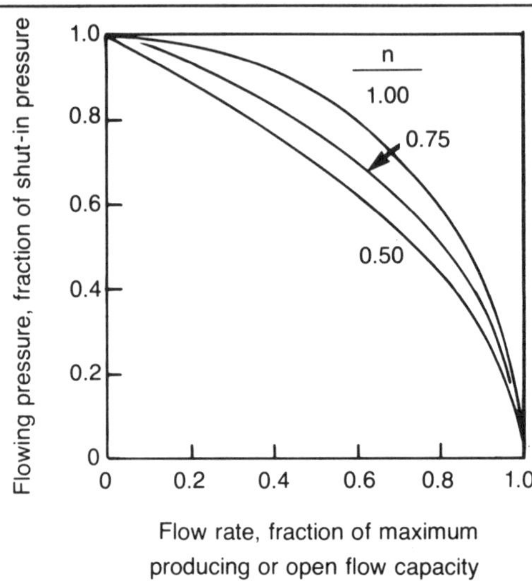

FIGURE 7–3 *Relationship Between Rate of Flow as a Ratio to Maximum Producing Capacity and Flowing Pressure as a Ratio to Shutin Pressure for a Gas Well*

shutin pressure at the wellhead of 3,100 psia and the exponent, n, is 0.75. The current pressure ratio is 1,000/3,100 = 0.323, and the corresponding ratio of 150 Mcfd to the maximum producing capacity is 0.92 from figure 7–3 or equation 7–9. One can then infer from this that the maximum producing capacity is 163 Mcfd. If the wellhead pressure is reduced to 500 psia as a result of reducing the pipeline pressure, the pressure ratio is 500/3,100 = 0.161, and the ratio of the rate of flow to the maximum producing capacity is 0.981. The new rate of flow would be 163 × 0.981 = 160 Mcfd. This corresponds to a gain in producing rate of 6.7% for a reduction in pipeline pressure of 50%. **Note:** The results would have been much different if the shutin wellhead pressure had been 1,100 psia instead of 3,100 psia. In fact, there would have been a 3.13-fold increase in rate of flow to 469 Mcfd, or 313% for a 50% reduction in pipeline pressure.

PRESSURE ADJUSTMENT OF OPEN-FLOW POTENTIALS.

At times, it may be desirable to adjust the open-flow potential of a well to a new shutin pressure. Such an adjustment is permissible at times when open-flow potentials are required periodically by regulatory bodies. The adjustment can be made readily by squaring the new pressure and reading the adjusted open-flow potential from the performance curve such as illustrated in figure 7–2, but calculation using an equation is preferable. If q_{of_1} is associated with $\bar{p}_{R_1}$, the new open-flow potential, q_{of_2}, can be calculated using the following equations:

$$q_{of_2} = C(\bar{p}_{R_2}{}^2)^n$$

and:

$$q_{of_1} = C(\bar{p}_{R_1}{}^2)^n$$

by dividing the two equations and solving them. The result is:

$$q_{of_2} = q_{of_1}\left(\frac{\bar{p}_{R_2}}{\bar{p}_{R_1}}\right)^{2n} \tag{7–10}$$

In the multipoint test example, the shutin pressure was 2,313.6 psia, the exponent, n, was 0.765, and the open-flow potential was 12,600 Mcfd. At a new shutin pressure of 2,050 psia, the corresponding open-flow potential is:

$$q_{of_2} = 12{,}600\left(\frac{2{,}050}{2{,}313.6}\right)^{1.53} = 12{,}600\ (0.831) = 10{,}471\ Mcfd$$

ONE-POINT TEST

The one-point test as used herein is intended to include deliverability testing and tests to determine the maximum allowable or the contractual maximum quantity for a given well. The test, usually with production into a gathering system, extends over a period of 72 hr, with the production during the last 24-hr period being taken as the test rate. If the rules of the regulatory body permit, the 72-hr test should start with a shutin well. The same precautions for the shutin time, measurement of shutin pressure, and preparation of the well for testing as were outlined previously should be observed for the one-point test. Regardless of the information required by the regulatory body or the gas sales contract, the engineer should use the one-point test as an opportunity to gain useful performance information on the well. For this purpose, record the shutin pressure to establish the pressure relationship with cumulative production. After flow is started, the wellhead flowing pressure and temperature and the instantaneous rates of flow should be taken at 0.5, 1, 2, 3, 6, 23, 24, 47, 48, 71, and 72 hr after starting the well flowing. Also, liquid production should be noted and measured if equipment is available. The rate of flow will be set according to the purpose of the test.

Deliverability Test

The deliverability of a gas well is defined as the capacity of the well to produce over a specified period of time against a given wellhead pressure called the *deliverability pressure*. Thus, the rate of flow or the deliverability of the well is determined by the capacity of the well to produce against the deliverability pressure. If the deliverability pressure should be more than the shutin pressure for the well, the well is said to have a deliverability of zero. For the rate of flow, it is seldom possible to set the flowing wellhead pressure exactly equal to the deliverability pressure. The deliverability, g_2, against the designated deliverability pressure, p_{tfd}, is calculated by the following:

$$q_d = q\left(\frac{\bar{p}_{ts}^{\,2} - p_{tfd}^{\,2}}{\bar{p}_{ts}^{\,2} - p_{tf}^{\,2}}\right)^n \tag{7-11}$$

where q is the observed producing rate. The value of the exponent, n, may be set at an arbitrary value by the regulatory body and, likewise, the deliverability pressure can be set at 80% of the average shutin pressure of a large number of wells in the same field or at 80% of the shutin pressure for the well. Also, it is possible to calculate a subsequent deliverability, q_{d1}, from a previous test or deliverability with the subsequent shutin pressure, p_{ts1}, and deliverability pressure, p_{tfd1}.

$$q_{d1} = q_d\left(\frac{\bar{p}_{ts1}^{\,2} - p_{tfd1}^{\,2}}{\bar{p}_{ts}^{\,2} - p_{tfd}^{\,2}}\right)^n \tag{7-12}$$

Tests for Maximum Take or Allowable

Frequently, it is necessary to file information with a regulatory body that shows a well is capable of producing an assigned allowable. Similar information may be required by a purchaser to show that a well is capable of producing at some factor (for example, 1.25) times the average contract quantity. Here, the principal precaution is to maintain the well in good condition to produce, if possible, more than the highest allowable to be assigned during the proration year or at a rate more than the factor times the average contract quantity. Unfortunately, not every well is capable of meeting these requirements. But it is the responsibility of the engineer to maintain the wells in the best condition and to see that tests of the highest quality are taken and reported.

ISOCHRONAL TEST AND EXAMPLE

The isochronal test is a series of one-point tests, each of which starts with the well shut in and the shutin pressure stablized or built up to 95% of the stabilized pressure. The precautions set out previously for preparing the well for testing should be observed. After flow is started in the well, flowing pressures, temperatures, and instantaneous rates of flow are measured at specified time intervals. Convenient time intervals are 0.5, 1, 2, 3, 6, 24 hr, etc. The isochronal test usually consists of 2 to 4 one-point tests with a minimum duration of 3 hr for each rate of flow. Occasionally, one of the series may be extended to 24 hr or more. Each shutin pressure should be at least 95% of the highest shutin pressure observed on the well during the series of one-point tests. The resultant data are presented on log-log coordinates in which the difference between the squares of the shutin pressure and the flowing pressure is the ordinate and the rate of flow is the abscissa. The performance curves are drawn to connect the points of equal times—hence the isochronal designation.

Isochronal tests can involve a great deal of shutin time for a well and are least inconvenient between the time the well is completed and the time it is connected to a pipeline. In this way, an isochronal test is run on a new well and subsequent testing would consist of one-point tests. Since the isochronal test determines the true value of the exponent *n* for the well, there should be no difficulty in correlating the subsequent one-point tests with the original isochronal test.

Data for an isochronal test of a gas well in the Hugoton field were given by Cullender and are set out in table 7–5 and illustrated in figure 7–4. The field data for an isochronal test can be recorded conveniently on the form illustrated by table 7–3 and the results calculated on the form shown on table 7–4. However, these forms are omitted for this test and the summary of the isochronal data is shown in table 7–5.

An examination of the isochronal test results in figure 7–4 shows that the data behave in a systematic way with time. The four curves are parallel for all practical purposes, and if each of the curves were extrapolated to obtain the open-flow potential, it is apparent that the potential of the well would diminish with time as the performance curve moved to the left. Also, if the well were started from shut-in to flow into a pipeline, the rate of flow would fall off rapidly. Fortunately, Cullender gives the results of 10 annual one-point

TABLE 7–5.

Isochronal Performance for a Gas Well in the Hugoton field, taken from Cullender (Table 4, copyrighted 1955, Society of Petroleum Engineers of AIME, used by permission.[3])

Date	Shutin Pressure (psia)	Duration of Flow (hrs)	Rate of Flow (Mcfd)	$\bar{p}_R^2 - p_{wf}^2$ (thousands)
12/03/51	352.4	0.5	983	5.37
		1.0	977	6.96
		2.0	970	8.93
		3.0	965	10.19
12/04/51	352.3	0.5	2,631	15.54
		1.0	2,588	19.82
		2.0	2,533	24.63
		3.0	2,500	27.72
12/05/51	351.0	0.5	3,654	21.63
		1.0	3,565	27.40
		2.0	3,453	34.03
		3.0	3,390	37.97
12/06/51	349.5	0.5	4,782	28.84
		1.0	4,625	35.96
		2.0	4,438	43.98
		3.0	4,318	48.96

tests (annual production tests) starting from shutin with performance data taken at the end of a flow period of 71.75 to 72.5 hr. These annual test data, taken over a period of 8 years, have been added to figure 7–4. Applying the method of least squares to the isochronal performance curves in figure 7–4 gave the values of the coefficient $C(t)$ and the exponent n shown in table 7–6. Also, the rates of flow expressed in Mcfd for the various times at $(\bar{p}_R^2 - p_{wf}^2) = 20.00$ thousands are given in the fourth column of table 7–6.

Table 7–6 shows that the coefficient, $C(t)$, in equation 7–2:

$$q = C(t)(\bar{p}_R^2 - p_{wf}^2)^n \qquad (7\text{–}2)$$

is definitely a function of time as was discussed previously in this chapter. Although the values for the exponent, n, show a slight increase from the 0.5 to the 3.0-hr curve, and the average value of 0.948 was used for the 72-hr curve, the general belief is that the n value is constant for the isochronal curves. The author knows of no data or proof to confirm the belief that n is constant. Since

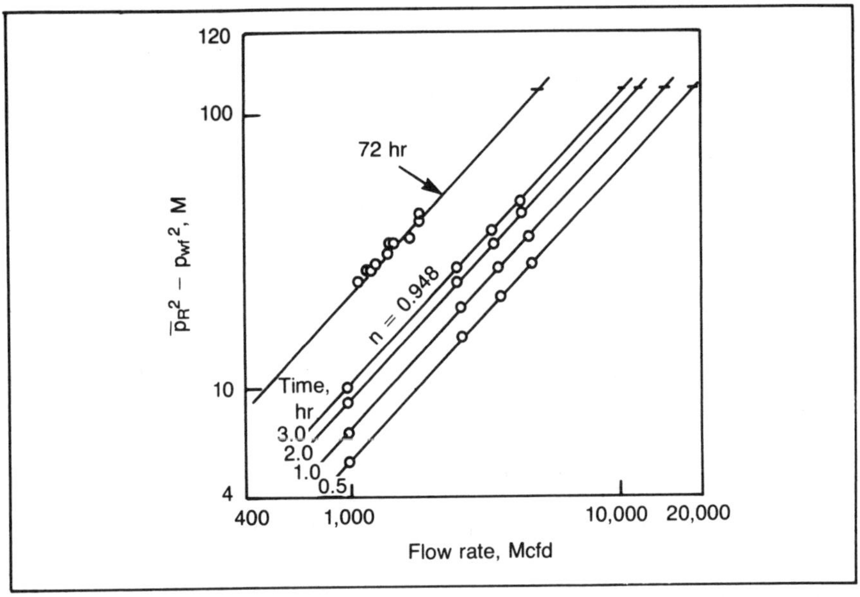

FIGURE 7–4 *Isochronal Test of Table 7–5 for a Gas Well in the Hugoton Field (From Cullender, Table 4, copyrighted 1955, Society of Petroleum Engineers of AIME, used by permission)*[3]

TABLE 7–6.
Values of the Coefficient, C(t), and Exponent, n, for Each of the Isochronal Performance Curves shown in figure 7–4 and for Rate of Flow, q_k, at $(\bar{p}_R^2 - p_{wf}^2)$ = 20.00 (thousands)

Time (hr)	*Coef-ficient [C(t)]*	*Expo-nent (n)*	*Flow rate $\Delta P^2 = 20.00$ (Mcfd)*
0.5	201.2	0.9414	3,376
1.0	155.3	0.9458	2,641
2.0	120.4	0.9522	2,087
3.0	105.1	0,9548	1,836
	Average	0.948	
72.0	52,02	0.948	890

the coefficient, $C(t)$, is the value of the rate of flow expressed in Mcfd at $(\bar{p}_R^2 - p_{wf}^2) = 1.00$ (thousands), q_k at $(\bar{p}_R^2 - p_{wf}^2) = 20.00$ follows the same rate of change with time. If the well in the isochronal test in table 7–5 and figure 7–4 is set to produce at a constant back pressure starting from shutin so that the difference of pressure squared is 20.00 (thousands), and the producing rate varies with time as shown in the fourth column of table 7–6 and illustrated by figure 7–5. The open circles are data points derived from table 7–6, and the crosses are from data on the same well that was presented by Cullender.

The data in figure 7–5 show that the gas well in the Hugoton field (Cullender's well No. 3) did not reach stabilization after 214 hr of flow because the coefficient, *C(t)*, was still a variable with time. At this point, the reader is justified to inquire as to the time required for stabilization for this particular gas well. This time becomes a subject for chapter 8 as mentioned previously.

Comments on Isochronal Testing

The advantage of isochronal testing gas wells is that it provides a method for eliminating the complicated pressure gradients in the reservoir that so often confuse the results of multipoint tests. It thereby permits the determination of the true value of the exponent, *n*, for the performance curves. The results of an isochronal test are good measures of well performance over a period of years. For example the results of the 72-hr production test (one-point test) shown in figure 7–4 indicate that the performance of the well was constant for a period of about 10 yr. This shows that the well had

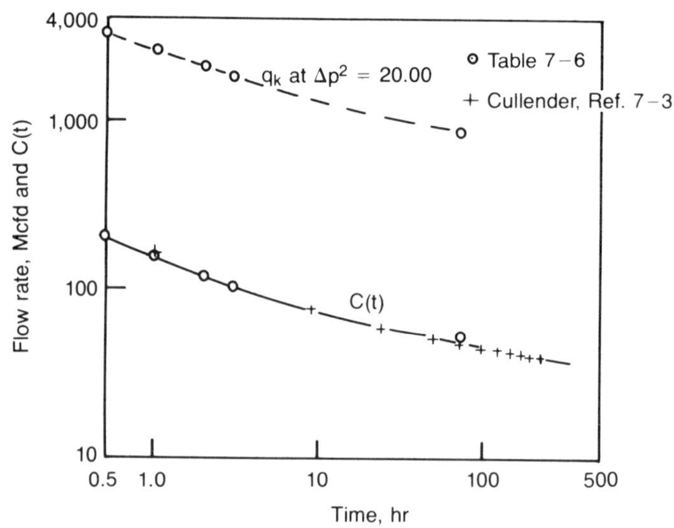

FIGURE 7–5 *Variation of Rate of Flow, Mcfd, and Coefficient,* C(t), *for Test Well in Table 7–6*

been maintained in its original operating condition. The principal difficulty with the isochronal test is the amount of time required for the shutin pressure to build between the flow rates imposed upon the well.

INFORMATION TEST

The results of an information test on an unconnected gas well is given in table 7–7 and illustrated as an isochronal test in figure 7–6. The well was

TABLE 7–7.
Example of an Information Test Including Pressure Buildup on a Gas Well Not Connected to a Pipeline

Elapsed Time (hr)	*Wellhead Pressure, $\bar{p}_{ts}$ or p_{tf} (psia)*	*Bottom-Hole Pressure, $\bar{p}_R$ or p_{wp} (psia)*	$\bar{p}_R^2 - p_{wf}^2$ *(thousands)*	*Rate of Flow (Mcfd)*
0.0	1,894	2,114.0	0.0	0
0.25	1,603	1,785.9	1,279.3	2,829
0.50	1,453	1,616.7	1,855.0	2,509
0.75	1,342	1,491.7	2,243.8	2,290
1.0	1,260	1,399.4	2,510.6	2,136
2.0	1,079	1,196.1	3,038.2	1,791
3.0	1,009	1,117.6	3,219.9	1.670
4.0	969	1,072.7	3,318.2	1,589
5.0	937	1,036.9	3,393.8	1,534
6.0	910	1,006.6	3,455.5	1,487
24.0	655	722.2	3,947.3	1,038
48.0	584	643.3	4,055.0	921
72.0	554	610.1	4,096.7	867
96.0	530	583.5	4,128.4	828
120.0	516	567.9	4,146.3	805
Well shut in				
0.25	661	728.8	—	0
0.50	780	861.3	—	0
0.75	884	977.4	—	0
1.0	975	1,079.3	—	0
2.0	1,242	1,378.9	—	0
3.0	1,405	1,562.4	—	0
4.0	1,497	1,666.1	—	0
5.0	1,554	1,730.4	—	0
6.0	1,592	1,773.2	—	0
24.0	1,757	1,959.4	—	0
48.0	1,803	2,011.3	—	0
72.0	1,826	2,037.3	—	0

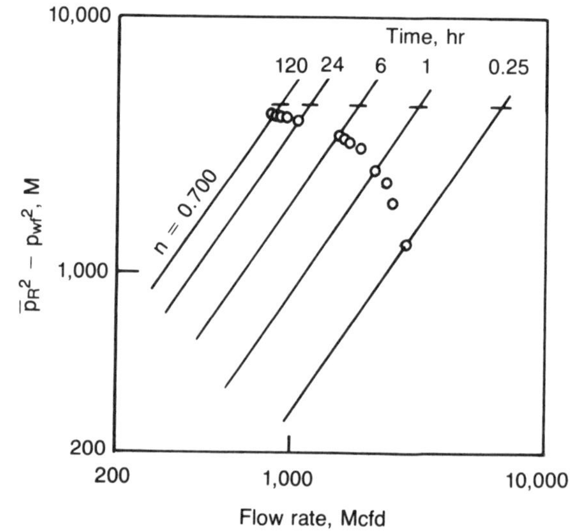

FIGURE 7–6 *Well Test Data from Table 7–7 Plotted as Isochronal Test with Arbitrary Exponent of 0.700 Assigned*

opened to flow to the atmosphere through a ¼-in orifice in a 2-in critical flow prover and was allowed to flow for 120 hr. Then it was shutin, and build-up pressures were taken for 72 hr. The well was not equipped with tubing. The well produced through 4,490 ft of 5½-in. casing with an ID of 5.012 in. The bottom-hole temperature was 95°F, and the surface temperature was about 60°F. An inspection of the results of the test shown in figure 7–6 shows that the well stabilized very slowly and that stabilization was not reached after 120 hr of flow. If an arbitrary exponent, n, of 0.700 is assigned to the data in figure 7–6, the open-flow potential decreases from 6,600 Mcfd after 0.25 hr of flow to 860 Mcfd after 120 hr of flow. Although the arbitrary assignment of the 0.700 exponent may have caused considerable error in the extrapolated open-flow potential of 6,600 Mcfd after 0.25 hr of flow, it caused little error in the potential of 860 Mcfd after 120 hr of flow. The well produced at a very large fraction of its open-flow potential after it had been flowing for 24 hr.

In the course of the test, the well was estimated to have produced a total of about 5,010 Mcf, but the average pressure in the casing was reduced from 2,004 to 542 psia. Thus, gas stored in the casing and in the open hole below the casing was reduced by an estimated 80 Mcf, and the formation and fractures must have given up about 4,930 Mcf during testing. The decrease in gas storage in the wellbore, although significant, was not a dominant factor but was only a contributing factor affecting the test results on the well.

The well can be called a small-capacity well, and one can conclude that the capacity of the well to produce into a pipeline was somewhat less than 800 Mcfd if the pressure in the pipeline was about 600 psia. Further, one must conclude that the well was still in unsteady-state flow, even after 120 hr. Further, one must have an understanding of unsteady-state flow behavior of gas wells, which is the subject of chapter 8.

The test given in table 7–7 was introduced under well testing as an example of information testing. **Note:** the size of the orifice plate in the critical flow prover was not changed during the test, and the working wellhead pressure and rate of flow did not change nearly so rapidly after the first 24 hr of the test although both decreased throughout the test. Now, the crucial question is at what point does the decrease in rate of flow and flowing pressure cease to be a result of unsteady-state flow and become a result only of the decrease in reserves for the well? In other words, when does the radius of pressure disturbance or the radius of investigation reach the outer boundary of the reservoir? These matters will be discussed in the next chapter.

SUMMARY

Flow testing gas wells falls into three and, at most, four types of testing. The first and oldest type of testing is the multipoint test in which three or four rates of flow are imposed upon a gas well without shutin periods between the rates of flow. Usually, the well is shut in until the pressure at the wellhead reaches a maximum before the first rate of flow is imposed upon the well. Although the multipoint test has been applied indiscriminately to all types of wells and reservoirs, the most meaningful information concerning the capacity of a well to produce gas is obtained when the radius of investigation reaches the reservoir boundary within the period of time selected for each rate of flow. The literature contains several methods for unscrambling the unsteady-state effects from a multipoint test, but the effort seldom seems worth the results.

Because wells were tested in which the radius of investigation did not reach the reservoir boundary within a relatively short period of time, one-point and isochronal tests were devised and used. The one-point test was used principally to replace the open-flow potential test in regard to proration formulas with a deliverability rate of flow. For this use, the one-point or deliverability test became a test run in which flow time period, flow rate, and flowing pressures were held at specified values or ranges of values, and the results were corrected to a standard working wellhead pressure at a specified time or narrow range of time. For this test, the well is usually shut in and the pressure is allowed to build to near its maximum value prior to imposing the rate of flow on the well.

The isochronal test is a series of one-point tests in which the well is shut in between each rate of flow. The shutin period as originally recommended by Cullender should be long enough to permit the pressure to return to a pressure comparable to that existing before the well was first opened. Good practice in the Hugoton fields has been to allow the pressure to return to at least 95% of the shutin pressure before the well was first opened to flow. The isochronal test method permits a determination of the true exponent of the performance curve for a particular gas well. Also, it permits the determination of the variation of the rate of flow from a well with time for a given pressure differential or the variation of the performance coefficient, $C(t)$, with time. The principal disadvantage is the lengthy shutin periods between the various rates of flow that may be required for some wells.

The information test is designed specifically to determine the unsteady-state behavior of a well and to determine at which point in time the unsteady-state behavior ends and pressure depletion of the reservoir begins. Ideally, the well should begin producing unrestrictedly from a shutin condition against a constant back pressure until the transition from unsteady-state flow to pressure depletion occurs. Ideal conditions are most closely followed in practice when the well is first connected to a pipeline that has a nearly constant pressure. In the case of an unconnected well (as was the case in table 7–7), the well should be opened to flow through a properly sized orifice and no further adjustments should be made in the rate of flow during the test. In either case, flowing pressures, temperatures, and rates of flow are measured at time intervals to give an even spacing of points when the data are plotted as in figure 7–5. Interpretation will be discussed in the following chapter.

PROBLEMS

1. Given a well with a wellbore radius of 0.25 ft in the center of a cylindrical reservoir with a radius of 2,640 ft in which the pressure is maintained constant at the outer edge of the cylinder, other values are as follows:

k = 500 millidarcies
h = 20 ft
$\bar{\mu}$ = 0.02 cp
$T_f = 460 + 120 = 580$
$\bar{z} = 0.90$

Calculate a multipoint test using equation 6–7 from chapter 6, plot the data on log-log paper, and determine the exponent, n, and the coefficient, C.

2. Assume that drilling mud infiltration has damaged the reservoir to a radius of 2 ft so that the permeability has become 50 md in the damaged portion. Recalculate the performance curve of problem 1 above. To what extent was the performance of the hypothetical well damaged? **Note:** Review equations 6–12 and 6–13 in chapter 6.

3. At a flow rate of 5,000 Mcfd, what percentage of the total pressure drop occurred out to the radius of 2 ft in the undamaged case and in the damaged case? **Note:** Assume that p_e^2 = 1068.36 thousands or p_e = 1033.6 psia.

4. Given the following test data on a gas well, construct a performance curve and write the equation of the curve.

Rate of Flow (Mcfd)	$\bar{p}_R^2 - \bar{p}_{wf}^2$ *(thousands)*
14.0	6.5
60.0	44.0
210.0	230.0

5. With $\bar{p}_R$ = 679 psia and using the following test data, construct and write the equation for the performance curve and determine the open-flow potential and rate of flow into a gathering line at 50 psia.

q_k	p_{wf}
6	671
14	656
40	583
80	400

Answer: $q_k = 0.96(\bar{p}_R^2 - p_{wf}^2)^{0.777}$

Open-flow potential = 112.7 Mcfd and q_k into line = 112 Mcfd.

6. A well produces separator gas with a specific gravity, γ_g, of 0.650 and hydrocarbon liquid with a gravity of 61.0° API at a ratio of 159,600 cu ft/bbl. Calculate the specific gravity of the flowing fluid. **Note:** See chapter 2.

Answer: $\gamma_{gf} = 0.668$

7. Cullender gives the following isochronal test data on Gas Well No. 1. (Table 1, copyrighted 1955, Society of Petroleum Engineers of AIME, used by permission.[3])

Date	$\bar{p}_R$	Duration of Flow (hr)	q_k	$\bar{p}_R^2 - p_{wf}^2$
10/ 3/44	435.2	24.0	9,900	97.70
10/24/44	436.8	1.0	4,656	22.52
—	—	3.0	4,587	27.26
—	—	23.0	4,440	38.67
12/11/45	394.7	0.1	2,016	4.61
		0.2	2,009	5.68
		0.5	2,001	7.10
		1.0	1,994	8.26
		3.0	1,980	10.38
		24.0	1,947	14.56
1/11/46	390.7	0.1	2,991	7.64
		0.2	2,977	9.21
		0.5	2,956	11.56
		1.0	2,941	13.28
1/15/46	394.0	0.2	7,327	26,09
		0.5	7,199	31.25
		1.0	7,092	35.52
		3.0	6,887	42.99
1/17/46	391.1	0.5	2,952	11.23
		1.0	2,937	12.87
		3.0	2,905	15.72
12/ 5/46	381.0	0.1	4,158	10.61
		0.2	4,130	12.72
		0.5	4,086	15.91
		1.0	4,052	18.40
		3.0	3,989	22.17

Construct isochronal performance curves and calculate the rate of flow from gas well no. 1 into a pipeline that would reduce the bottom-hole pressure, p_{wf}, to 366.7 psia. Delivery rates should be calculated after 0.1, 0.2, 0.5, 1.0, 3.0, and 24.0 hr of flow to the pipeline. Assume the well was opened to flow on December 5, 1946. **Note:** Save your results to work a problem in chapter 8.

8. Use a selected rate of flow from the table in problem 7. Calculate p_{wf}^2 and plot the answer on semilog paper with time on the log scale. Extrapolate to 72 and 240 hr and estimate production rates.

9. Figure 2 of reference 3 lists multipoint tests of 24-hr duration on gas well no. 1 for each rate of flow from which the following coefficients, *C*, and exponents, *n*, can be calculated. (Figure 2, copyrighted 1955, Society of Petroleum Engineers of AIME, used by permission.[3])

Curve	*n*	*C*	*Description*
A	1.097	66	Reverse sequence
B	0.701	336	Normal sequence
C	0.776	234	Normal sequence
D	0.867	190	Isochronal test

Plot the performance curves and decide whether a production rate could have been calculated from the multipoint data. **Note:** The difference in the multipoint tests is due to the difference in rates of flow and the order in which the rates of flow were imposed on the well.

10. Calculate the absolute open-flow potentials from the isochronal data for well no. 1, problem 7, for the various periods of time. Which potential is more nearly representative of the capacity of the well to produce? If you, as an engineer, had an opportunity to purchase well no. 1, which type of test data (multipoint or isochronal) would be most helpful?

REFERENCES

1. H.R. Pierce and E.L. Rawlins. *The Study of a Fundamental Basis for Controlling and Gaging Natural Gas Wells; Part 2—A Fundamental Relation for Gaging Gas Well Capacities.* Bureau of Mines Report of Investigations 2930, 1929.
2. E.L. Rawlins and M.A. Schellhardt. *Back-Pressure Data on Natural Gas Wells and Their Application to Production Practices.* Bureau of Mines Monograph 7, 1935.
3. M.H. Cullender. "The Isochronal Performance Method of Determining the Flow Characteristics of Gas Wells." *Trans. AIME* 204 (1955), p. 137.
4. Interstate Oil Compact Commission. *Manual of Back-Pressure Testing of Gas Wells.* Oklahoma City, Oklahoma, 1962.
5. Kansas State Corporation Commission. *Manual of Back-Pressure Testing of Gas Wells.* Topeka, Kansas, 1959.

6. Oklahoma Corporation Commission. *Manual of Back-Pressure Testing of Gas Wells.* Oklahoma City, Oklahoma, 1966.
7. Oil and Gas Conservation Board. *Theory and Practice of the Testing of Gas Wells.* Calgary, Alberta, Canada, 1964.
8. Energy Resources Conservation Baord. *Gas Well Testing—Theory and Practice*, 4th ed. Calgary, Alberta, Canada, 1979.
9. J.S. Aronofsky and R. Jenkins. "A Simplified Analysis of Unsteady Radial Gas Flow." *Trans. AIME* 201 (1954), 149.
10. R. Jenkins and J.S. Aronofsky. "Unsteady Radial Flow of Gas Through Porous Media." *Jour. Appl. Mech., ASME* 20 (June 1953).
11. G.H. Bruce, D.W. Peaceman. H.H. Rachford, and J.D. Rice. "Calculation of Unsteady-State Gas Flow Through Porous Media." *Trans. AIME* 198 (1953), p. 79.
12. R.V. Smith. "Unsteady-State Gas Flow into Gas Wells." *Trans. AIME* 222 (1961), p. 1,151.
13. A.F. van Everdingen. "The Skin Effect and Its Influence on the Productive Capacity of a Well." *Trans. AIME* 198 (1953), p. 171.
14. R. Al-Hussainy, H.J. Ramey Jr., and P.B. Crawford. "The Flow of Real Gases Through Porous Media." *Trans. AIME* 237 (1966), p. 624.
15. C.S. Matthews and D.G. Russell. *Pressure Buildup and Flow Tests in Wells.* vol. 1. Monograph Series, Dallas: SPE of AIME (1967).
16. R.C. Earlougher Jr. *Advances in Well Test Analysis.* vol. 5. Monograph Series. Dallas: SPE of AIME (1977).

8 Unsteady-State Flow Behavior of Gas Wells

In the preceding chapter, conventional methods based on steady-state behavior were inadequate for testing and analyzing test results when a well exhibited any marked degree of unsteady-state flow behavior. The unsteady-state flow behavior of water wells and oil wells has received an immense amount of study since the first papers on unsteady flow in oil reservoirs were published in the 1930s. The work of van Everdingen and Hurst and that of Miller, Dyes, and Hutchinson on flow in reservoirs[1,2] and of Horner on pressure buildup in wells[3] started an avalanche of papers on unsteady-state flow that continues until this day. The work of Aronofsky and Jenkins; Jenkins and Aronofsky; Bruce, Peacemen, Rachford, and Rice; and Smith showed that the numerical solutions for transient flow of gases in porous media could be approximated by the solutions for transient liquid flow.[4,5,6,7]

Later, Al-Hussainy, Ramey, and Crawford introduced the concept of the real gas potential, $m(p)$, which has the dimensions of (pressure)^2/viscosity.[8] At times, the real gas potential has been called real gas pseudopressure and pseudopressure. It is defined:

$$m(p) = 2 \int_{p_m}^{p} \frac{p}{\mu(p) z(p)} \, dp \qquad (8\text{–}1)$$

where a symbol followed by (p) indicates that the symbol is a function of pressure, p_m is a low base pressure, and the flow is considered to be isothermal. Under isothermal conditions, $\mu(p)$ and $z(p)$ are functions of pressure alone. The real gas potential $m(p)$ can be substituted for pressure in liquid solutions for unsteady-state flow in porous media.

It is beyond the scope of this book to deal with the fundamentals of unsteady-state flow of gases. Therefore, refer to a comprehensive review of the subject and to pertinent literature presented by the Energy Resources Conservation Board of Alberta, Canada.[9]

Before further discussing the important aspects of unsteady-state gas flow, it will be helpful to ask what the practicing engineer requires of gas well testing and what questions his management expects his gas well testing to answer. Returning to the example in chapter 7 of the information test on a gas well given in table 7–7 and figure 7–6, a gas well produces at a rate of 2,829 Mcfd after 0.25 hr of flow although the rate had decreased to 805 Mcfd after 120 hr of flow. Keeping in mind that the well was not connected to a pipeline, the primary questions are:

1. If connected to a pipeline is it possible to make a reasonable forecast of production?
2. Will connecting the well to a pipeline be profitable?
3. Assuming the geology is favorable, do the reserves justify drilling a second well to the same reservoir?

In this case, every question hinges on the production forecast. The same questions can be asked in a different way—was the decrease in producing rate over the period of 120 hr principally caused by depletion of the reserve or was it caused by unsteady-state effects?

To some extent, the question about reserves can be answered by comparing the pressure before the test to the pressure observed after the 72-hr buildup. The difference of 76.7 psi between the before and after pressures caused by the production of 5,010 Mcf of gas indicates that the gas in place after the test was at least:

$$\frac{2{,}037.3}{76.7} \times 5{,}010 = 133{,}100 \ Mcf$$

This calculation assumes that the pressures given were representative of stabilized pressures in the reservoir. Note also that the assumption was made that the gas in the reservoir acted as an ideal gas. Unfortunately, one is left with a vague idea of the reserves and without a production forecast. A good production forecast is a necessity for an economic analysis.

In posing the primary questions about reserves and profitability, thus far information gained from testing concerning the characteristics of the reservoir and the efficiency of the well completion has been ignored. These are equally important to the reservoir engineer and will be treated in later discussions.

Referring to figure 7–6, note that, as time went on, the well produced closer to its current open-flow potential. If the well had been producing into a pipeline with a constant pressure of 600 psia, the well would have been "floating" on the line after little more than 24 hr. In other words, the wellhead pressure would have become constant at 600 psia, and the rate of flow would have been the only variable. This type of operation, production against a constant back pressure, is a common mode of operation for small-capacity gas wells. Therefore, consider the constant inner boundary pressure case as the principal interest in considering unsteady-state effects.

BEHAVIOR AGAINST A CONSTANT WELLBORE PRESSURE

Solutions to the unsteady-state theory for predicting decreasing producing rates against constant wellbore pressures were published for liquid (slightly compressible), single-phase systems by Moore, Schilthuis, and Hurst and by Hurst.[10,11] The results were presented in graphical form for radial systems in terms of a dimensionless flow rate, q_D, and a dimensionless time, t_D. Tabular values of q_D and t_D are found in reference 12 for the infinite reservoir and in references 13 and 14 for finite reservoirs. The dimensionless flow rate, q_D, and dimensionless time, t_D, are defined as follows:

$$q_D = \frac{141.3\ q_o(t)\mu B}{kh(p_i - p_{wf})} \qquad (oil) \tag{8–2}$$

$$t_D = \frac{0.00634\ kt}{\phi\mu c_t r_w^2} \tag{8–3}$$

where $q_o(t)$ is the rate of oil production (b/d), t is time (days), and c_t is the total compressibility (psi^{-1}). Fetkovich and Fetkovich & Thrasher have collected the tabular values and prepared convenient charts of q_D as a function of t_D for plane radial systems with infinite and finite outer boundaries and constant pressure at the inner boundary.[15,16] The charts of Fetkovich and Fetkovich & Thrasher are reproduced from tabular values in figures 8–1A and 8–1B where q_D varies in the range of 0.001–1.00 and t_D varies from 0.0001–10^9. In addition to the tabular values for plane radial systems, Fetkovich and Thrasher have added the constant pressure solution for a vertical fracture with infinite conductivity given by Agarwal, Carter, and Pollock, which is shown as a dashed line on figure 8–1A.[17]

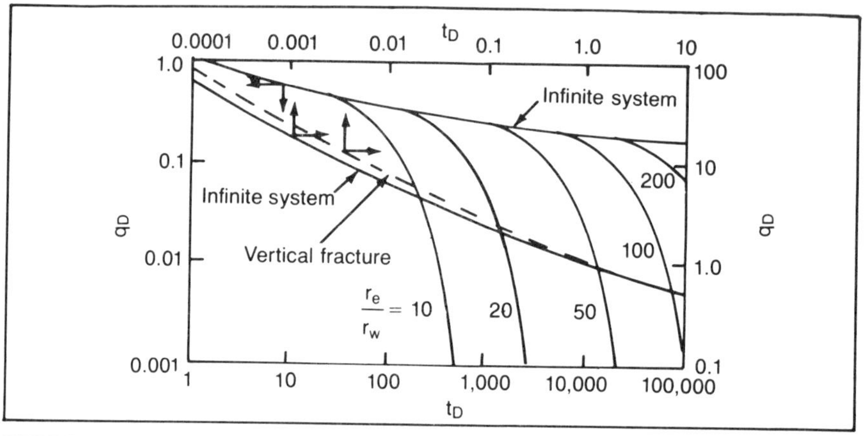

FIGURE 8–1A *Dimensionless Flow Rate (q_D) and Time Function (t_D) for Plane Radial Systems with Infinite and Finite Outer Boundaries and Constant Pressure at the Inner Boundary, After Fetkovich and Fetkovich & Thrasher[15,16]*

It is the intention here to use the charts illustrated on figures 8–1A and 8–1B for type curve matching, but because they are reproduced, they are much too small.* However, selected numerical values obtained for q_D as a function of t_D by Ferris et al. for a plane radial system in an infinite isotropic porous medium with constant pressure at the inner boundary are given in table 8–1. The constant pressure solution for a plane radial system with an infinite conductivity vertical fracture taken by Fetkovich and Thrasher from Agarwal, Carter, and Pollock was shown as a dashed line on figure 8–1A. Values of q_D and corresponding t_D values were read from the dashed curve on a plot much larger than that in figure 8–1A and are given in table 8–2. In a similar fashion, the values of q_D and corresponding t_D values were obtained for table 8–3. It must be emphasized that the values given in tables 8–2 and 8–3 are not numerical solutions to the appropriate flow functions but were read from large plots of the numerical solutions. The values in tables 8–2 and 8–3 are included here for convenience to enable the reader to construct duplicates of figures 8–1A and 8–1B.

* The full-size type curves published by Fetkovich with a grid suitable for actual use are available on written request from SPE Book Order Department, 6200 N. Central Expressway, Dallas, TX 75206. Specify SPE 9086 and include $3.00 prepayment for each order of "Type Curves for Decline Curve Analysis Using Type Curves."

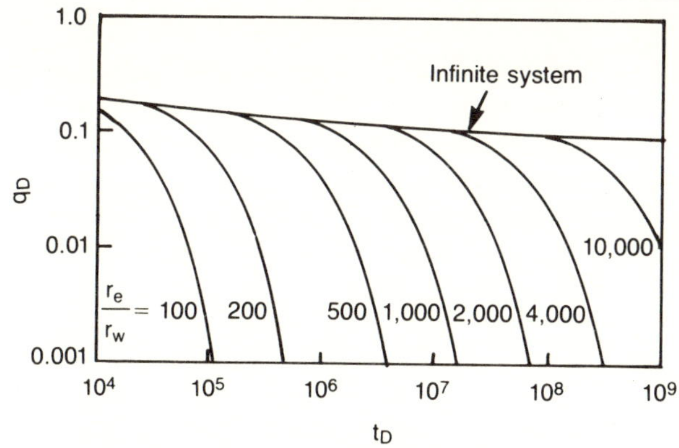

FIGURE 8–1B *Dimensionless Flow Rate (q_D) and Time Function (t_D) for Plane Radial Systems with Infinite and Finite Outer Boundaries and Constant Pressure at the Inner Boundary, After Fetkovich*[15]

TABLE 8–1.

Values of q_D as a Function of t_D for a Plane Radial System, Infinite Outer Boundary with Constant Pressure at the Inner Boundary

	t_D						
Multiplier	10^{-4}	10^{-3}	10^{-2}	10^{-1}	1	10	10^2
1	56.9	18.34	6.13	2.249	0.985	0.534	0.346
2	40.4	13.11	4.47	1.716	0.803	0.461	0.311
4	28.7	9.41	3.30	1.333	0.667	0.405	0.283
6	23.5	7.77	2.78	1.160	0.602	0.377	0.268
8	20.4	6.79	2.46	1.057	0.562	0.359	0.258
10	18.3	6.13	2.25	0.985	0.534	0.346	0.251

	t_D						
Multiplier	10^3	10^4	10^5	10^6	10^7	10^8	10^9
1	0.251	0.1964	0.1608	0.1360	0.1177	0.1037	0.0927
2	0.232	0.1841	0.1524	0.1299	0.1131	0.1002	0.0899
4	0.215	0.1733	0.1449	0.1244	0.1089	0.0968	0.0872
6	0.206	0.1675	0.1408	0.1213	0.1066	0.0950	0.0857
8	0.200	0.1636	0.1380	0.1192	0.1049	0.0937	0.0846
10	0.196	0.1608	0.1360	0.1177	0.1037	0.0927	0.0838

Source: Ferris et al., *Theory of Aquifer Tests.*

TABLE 8–2.
Values of q_D as a Function of t_D for a Plane Radial System with an Infinite Conductivity Vertical Fracture, Infinite Outer Boundary and Constant Pressure at the Inner Boundary

	t_D				
Multiplier	10^{-4}	10^{-3}	10^{-2}	10^{-1}	1
1	76.0	23.8	7.9	2.71	1.08
2	53.0	17.1	5.7	1.98	0.86
4	37.5	12.2	4.15	1.52	0.69
6	30.8	10.0	3.42	1.30	0.61
8	26.5	8.8	3.00	1.18	0.56
10	23.8	7.9	2.71	1.08	0.53

Note: the above values were read from charts and are not numerical solutions of the flow functions.
Source: Fetkovich and Thrasher, "Constant Well Pressure Testing and Analysis in Low Permeability Reservoirs."

Perhaps a more-thorough understanding of unsteady-state flow can be gained by discussing the significance of figures 8–1A and 8–1B. If flow is started in a well against a constant back pressure in a plane radial flow regime from a reservoir that is in a steady state of pressure equilibrium, the variation of the flow rate with time on a dimensionless basis will follow the lower solid line in figure 8–1A at the lesser values of dimensionless time. If the well should have a vertical fracture of infinite conductivity, the variation of the flow rate with time will follow the dashed curve in figure 8–1A. In either case, the reservoir with respect to flow will act as if it were an infinite system. In other words, the pressure disturbance caused by starting the well flowing will not have reached the boundary of the reservoir. Flow in the reservoir with the fracture will become identical to that in the reservoir without a fracture at dimensionless time, t_D, between 1 and 10. If the reservoir boundary is at $r_e/r_w = 1{,}000$, the variation of q_D with t_D will continue as if the reservoir were of infinite extent until t_D is almost 10^6 (see figure 8–1B) and then will follow the stem labelled 1,000. At t_D somewhat over 10^7, the value of q_D will be 0.001. Since the curves in figures 8–1A and 8–1B are dimensionless, the solutions to the flow functions are general for a plane radial system.

Conversion to Gas Terminology

Equation 8–2, as given previously, is expressed in terms of barrels per day, and after multiplication by the formation volume factor, B, the flow rate becomes reservoir barrels per day. For flow of liquids, the volume of a unit

TABLE 8–3.

Values of q_D as a Function of t_D for a Plane Radial System with r_e/r_w and Constant Pressure at the Inner Boundary[15]

	r_e/r_w									
q_D	*10*	*20*	*50*	*100*	*200*	*500*	*1,000*	*2,000*	*4,000*	*10,000*
0.4	2.90×10^1	—	—	—	—	—	—	—	—	—
0.3	5.15×10^1	1.51×10^2	—	—	—	—	—	—	—	—
0.2	8.70×10^1	3.40×10^2	1.65×10^3	4.50×10^3	—	—	—	—	—	—
0.15	1.09×10^2	4.75×10^2	2.85×10^3	1.04×10^4	3.30×10^4	1.28×10^5	—	—	—	—
0.10	1.43×10^2	6.60×10^2	4.45×10^3	1.80×10^4	7.00×10^4	4.03×10^5	1.43×10^6	4.90×10^6	1.62×10^7	6.40×10^7
0.08	1.60×10^2	7.62×10^2	5.31×10^3	2.24×10^4	9.00×10^4	5.50×10^5	2.12×10^6	8.00×10^6	3.00×10^7	1.60×10^8
0.06	1.83×10^2	9.00×10^2	6.50×10^3	2.78×10^4	1.15×10^5	7.50×10^5	3.00×10^6	1.20×10^7	4.75×10^7	2.79×10^8
0.04	2.17×10^2	1.08×10^3	8.20×10^3	3.60×10^4	1.54×10^5	1.03×10^6	4.30×10^6	1.76×10^7	7.10×10^7	4.50×10^8
0.03	2.43×10^2	1.22×10^3	9.25×10^3	4.07×10^4	1.80×10^5	1.23×10^6	5.15×10^6	2.15×10^7	8.85×10^7	5.80×10^8
0.02	2.75×10^2	1.40×10^3	1.08×10^4	4.92×10^4	2.16×10^5	1.50×10^6	6.42×10^6	2.71×10^7	1.14×10^8	7.35×10^8
0.015	3.00×10^2	1.54×10^3	1.20×10^4	5.45×10^4	2.43×10^5	1.70×10^6	7.30×10^6	3.10×10^7	1.31×10^8	8.50×10^8
0.010	3.32×10^2	1.72×10^3	1.36×10^4	6.28×10^4	2.80×10^5	1.96×10^6	8.50×10^6	3.65×10^7	1.55×10^8	1.07×10^9
0.008	3.50×10^2	1.83×10^3	1.44×10^4	6.70×10^4	3.00×10^5	2.12×10^6	9.20×10^6	3.95×10^7	1.68×10^8	—
0.006	3.75×10^2	1.95×10^3	1.57×10^4	7.25×10^4	3.26×10^5	2.31×10^6	1.00×10^7	4.35×10^7	1.83×10^8	—
0.004	4.04×10^2	2.14×10^3	1.72×10^4	8.05×10^4	3.63×10^5	2.60×10^6	1.13×10^7	4.88×10^7	2.10×10^8	—
0.003	4.28×10^2	2.27×10^3	1.83×10^4	8.60×10^4	3.90×10^5	2.80×10^6	1.22×10^7	5.30×10^7	2.28×10^8	—
0.002	4.61×10^2	2.46×10^3	2.00×10^4	9.40×10^4	4.28×10^5	3.08×10^6	1.34×10^7	5.85×10^7	2.52×10^8	—
0.0015	4.85×10^2	2.60×10^3	2.12×10^4	1.00×10^5	4.55×10^5	3.26×10^6	1.43×10^7	6.25×10^7	2.71×10^8	—
0.0010	5.20×10^2	2.80×10^3	2.30×10^4	1.06×10^5	4.92×10^5	3.60×10^6	1.56×10^7	6.80×10^7	2.98×10^8	—

Note: the above values were read from charts and are not numerical solutions of the flow functions.

Source: Fetkovich, "Decline Curve Analysis Using Type Curves," *JPT*, June 1980.

weight of the liquid and the viscosity are relatively constant over the range of pressures found in oil wells and it is inconsequential to assume them to be constant. Usually for gas flow, it is necessary to assume that the volume flowing was constant at an average pressure. Also, it is expedient to express the volume in standard cubic feet. Thus:

$$q_o(t)B = \frac{178.1071(2)\ q_k p_b\ \bar{z}\ T}{(p_i + p_{wf})\ T_b} \tag{8–4}$$

and by substitution, equation 8–2 becomes

$$q_D = \frac{q_k p_b\ \bar{z}\bar{\mu}\ T}{19.87 \times 10^{-6} kh\ (p_i^2 - p_{wf}^2)\ T_b} \tag{8–5}$$

for gases. The equation defining the dimensionless time, t_D, for gases remains the same as in equation 8–3.

In the case of a vertical fracture, the dimensionless time based on the fracture half-lenght, X_f, according to Fetkovich and Thrasher is:

$$t_{DX_f} = \frac{0.00634\ k\ t}{\phi\ \mu\ C_t\ X_f^2} \tag{8–6}$$

The equivalent wellbore radius, r_w', is

$$r_w' = \frac{X_f}{2} \tag{8–7}$$

Therefore:

$$t_{Dr_{w'}} = 4t_{Dxf} \tag{8–8}$$

Since:

$$\frac{p_i^2 - p_{wf}^2}{\bar{\mu}\bar{z}} = m(p_i) - m(p_{wf}) \tag{8–9}$$

equation 8–5 becomes:

$$q_D = \frac{q_k\, p_b\, T}{19.87 \times 10^{-6}\, kh\, [m(p_i) - m(p_{wf})]\, T_b} \tag{8–10}$$

on the basis of the real gas potential, $m(p)$.

After the effective wellbore radius, r_w', is known, the van Everdingen skin factor, s, can be calculated by the following:[18]

$$s = -\ell n \left(\frac{r_w'}{r_w} \right) \tag{8–11}$$

or:

$$s = -2.303\, log \left(\frac{r_w'}{r_w} \right)$$

where r_w is the radius of the wellbore as drilled.

EXAMPLE AND ANALYSIS OF UNSTEADY-STATE BEHAVIOR

The test results given in table 7–7 and illustrated in figure 7–6 are typical of the unsteady-state behavior of a gas well. Also, they are representative of the type of testing and results encountered frequently by the natural gas engineer. The test, however, was not run at a constant back pressure, which makes it desirable to correct the test results to a constant back pressure, but, unfortunately, there has been no generally accepted method for the correction. Winestock and Colpitts normalized test results by dividing $(p_R^2 - p_{wf}^2)$ with and without a correction for turbulance by the rate of flow, q_k.[19] In this section, several approaches will be examined.

Referring to figure 7–6, note that as time went on the well produced closer and closer to its open-flow potential. If the well were producing into a pipeline with a constant pressure of about 600 psia, the well would have been floating on the line after little more than 24 hr. In other words, the wellhead pressure would have become constant at about 600 psia and the rate of flow would have become the only variable. If one assumes that the bottom-hole

TABLE 8–4.
Normalization of Test Results from Table 7–7 to Give the Rate of Flow, q_2, at a Constant Back Presure, p_{wf}, of 662 psia

Elapsed Time (hr)	*($\bar{p}_R^2 - p_{wf}^2$) (thousands)*	q_1	q_2 *at ($\bar{p}_R^2 - p_{wf}^2$) = 4,031 (thousands)*		
			n = 0.70	*n = 0.85*	*n = 1.00*
0.25	1,279.3	2,829	6,317	7,504	8,914
0.50	1,855.0	2,509	4,320	4,853	5,452
0.75	2,243.8	2,290	3,451	3,768	4,114
1.0	2,510.6	2,136	2,975	3,194	3,430
2.0	3,038.2	1,791	2,183	2,278	2,376
3.0	3,219.9	1,670	1,954	2,021	2,091
4.0	3,318.2	1,589	1,821	1,875	1,930
5.0	3,393.8	1,534	1,730	1,776	1,822
6.0	3,455.5	1,487	1,656	1,695	1,735
24.0	3,947.3	1,038	1,053	1,057	1,060
48.0	4,055.0	921	917	916	916
72.0	4,096.7	867	857	855	853
96.0	4,128.4	828	814	811	808
120.0	4,146.3	805	789	786	783

pressure, p_{wf}, was constant at 662 psia, it is possible to calculate the rate of flow for the well producing against the bottom-hole pressure, providing one also assumes values for the exponent, n, of 1.00, 0.85, and 0.70 for the isochronal type curve for the well. This is accomplished by the following equation:

$$q_2 = q_1 \left(\frac{\Delta p_2^{\,2}}{\Delta p_1^{\,2}} \right)^n \tag{8–12}$$

where $\Delta p^2 = \bar{p}_R^{\,2} - p_{wf}^{\,2}$. The results of the calculations are shown in table 8–4 and in figure 8–2.

TYPE CURVE MATCHING

In the preceding section, the purpose was to adjust the test results so that the results more nearly conformed to data that would have been obtained if the well had been producing against a constant back pressure at the sand face in the well. It is evident from figure 7–6, table 8–4, and figure 8–2 that the value of the exponent, n, selected for the normalization process makes very

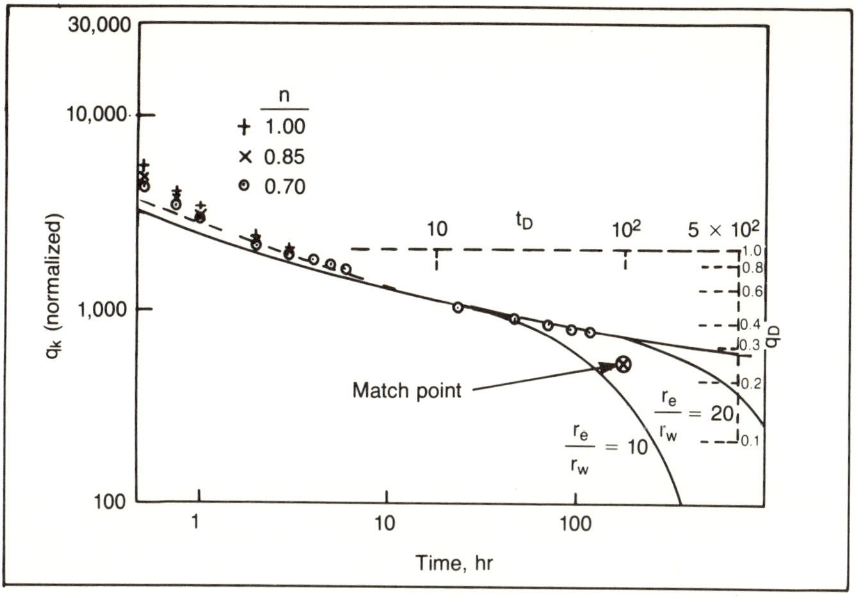

FIGURE 8–2 *Normalized Test Data from Table 8–4 Imposed on Flow Functions from Figures 8–1A and 8–1B; Pressures Squared Used*

little difference in the final results. After about 3 hr of test time, the normalized points converge very closely. For purposes of further analysis, the data will be considered identical.

The normalized data points are plotted on log-log tracing paper with 3-in. cycles as illustrated in figure 8–2. Tracing paper is then superimposed on copies of figures 8–1A and 8–1B on 3-in. cycle log-log paper and the tracing paper is moved back and forth, keeping the coordinates parallel, until the points on the tracing paper match the curves in figure 8–1A. It is possible to match only the points from 24 to 120 hr of the test as shown in figure 8–2. The underlying grid scale of figure 8–1A has been shown in part by the dashed lines in figure 8–2. An arbitrary match point is selected and is indicated in figure 8–2 by an X inside a circle. The match points at q_k = 535 Mcfd and time = 181 hr or 7.54 days on the tracing paper are equivalent to a dimensionless flow rate, q_D = 0.25, and a dimensionless time, t_D = 100. Rearranging equation 8–5 and setting the basis of the gas measurement to be 14.65 psia and 60°F, the result is:

$$kh = \frac{1{,}418\ q_k\ \bar{z}\bar{\mu}\ T}{q_D(\bar{p}_R^{\,2} - p_{wf}^{\,2})} \tag{8–13}$$

where $\bar{p}_R$ has been substituted for p_i. The value of kh can then be calculated:

$$kh = 1{,}418 \frac{(535)(0.863)(0.0140)(555)}{(0.25)(4{,}031{,}000)}$$

where:

$$kh = 5.05 \text{ md-ft}$$

or:

$$k = 5.05/31 = 0.16 \; md$$

Here, the average compressibility factor, $\bar{z}$, has been taken at the reservoir temperature of 95°F and the average of $p_R = 2{,}114$ and $p_{wf} = 662$. The average viscosity, $\bar{\mu}$, is obtained in a similar manner. The thickness, h, is estimated from electric logs to be 31 ft, which gives the permeability as 0.16 md, and the porosity capable of holding gas, $\phi(1 - S_w)$, is estimated to be 9% from electric logs.

If one rearranges equation 8–3 and uses the calculated permeability, 0.16, one can calculate the effective wellbore radius, r_w':

$$r_w' = \left(\frac{0.00634 \; k \; t}{\phi \; \mu \; C_t \; t_D} \right)^{0.5} \tag{8–14}$$

$$r_w' = \left[\frac{(0.00634)(0.16)(7.54)}{(0.09)(0.01607)(0.000455)(100)} \right]^{0.5}$$

$$r_w' = 11 \text{ ft}$$

The viscosity, μ, and the total compressibility, c_t, is taken at the original pressure in the reservoir because the original pressure is the condition of the reservoir through which the outer edge of the pressure disturbance moves. Since the compressibility, c, of the gas is so much more than that of the liquids and rock in the reservoir, the compressibility is for the gas alone. It is calculated by means of equation 2–15.

If the well's wellbore diameter as drilled is 6 in. or $r_w = 0.25$ ft, one can calculate the skin factor using equation 8–11.

$$s = -\ell n \frac{r_w'}{r_w} = -\ell n \frac{11}{0.25} = -3.78$$

The reader must be warned that the theory used in calculating the skin effect was worked for flow of liquids where non-Darcy effects or turbulance did not occur. However, for flow of gas there is a high probability of pressure losses caused by turbulance and any such losses would show up as a positive skin effect. In all probability, the calculated skin effect of −3.78 is too high—in other words, the true Hurst-van Everdingen skin effect would be a larger negative number.

Use of the Real Gas Potential, *m(p)*

Before discussing the analysis of the test data, one should use the real gas potential, $m(p)$, to analyze the test results of the previous section. First, calculate the real gas potential values over the pressure range of interest. Recalling the definition given in equation 8–1:

$$m(p) = 2 \int_{p_m}^{p} \frac{p}{\mu(p) z(p)} dp \qquad (8\text{–}1)$$

It was decided to use 15 psia for p_m and integrate from 15 to 50 psia, from 50 to 100 psia and from thereon in 100 psi increments to 2,200 psia. The actual steps in the numerical integration are not shown, but the values of interest are given in table 8–5. (Compressibility factors and viscosities are calculated by equations 2–11 and 2–17, respectively. The specific gravity of the gas is 0.560, the molecular weight is 16.22, and the reservoir temperature is 95°F.) The test results are normalized to an $m(\bar{p}_R) - m(p_{wf})$ value of 326.51×10^6 at exponent n values of 0.70, 0.85, and 1.00 and are illustrated in figure 8–3. Again, make the curve match and select a match point marked by an X in a circle. The match point at $q_k = 450$ Mcfd and time $t = 175$ hr or 7.29 days on the tracing paper are equivalent to a dimensionless flow rate, $q_D = 0.20$, and a dimensionless time, $t_D = 150$.

Rearranging equation 8–10 and setting 14.65 psia and 60°F as the basis of the gas measurement, the result is:

$$kh = \frac{1{,}418\, q_k T}{q_D\, [m(\bar{p}_R) - m(p_{wf})]} \qquad (8\text{–}15)$$

TABLE 8–5.
Values of the Real Gas Potential, m(p), for Pressures of Interest in Test Results on a Gas Well from Table 7–7. Temperature is 95°F.

Elapsed Time (hr)	q_k *(Mcfd)*	$\bar{p}_R$ *or* p_{wf}	*Compressibility factor (z)*	*Viscosity,* μ *(cp)*	$m(p) \times 10^6$	$m(\bar{p}_R) - m(p_{wf}) \times 10^6$
0.0	0	2,114.0	0.8305	0.01607	346.12	0.0
0.25	2,829	1,785.9	0.8402	0.01508	265.69	98.43
0.50	2,509	1,616.7	0.8482	0.01460	219.73	144.39
0.75	2,290	1,491.7	0.8555	0.01427	188.12	176.00
1.0	2,136	1,399.4	0.8615	0.01404	166.16	197.96
2.0	1,791	1,196.1	0.8766	0.01356	122.15	241.97
3.0	1,670	1,117.6	0.8830	0.01339	106.84	257.28
4.0	1,589	1,072.7	0.8869	0.01330	98.54	265.58
5.0	1,534	1,036.9	0.8900	0.01322	92.10	272.02
6.0	1,487	1,006.6	0.8927	0.01316	86.83	277.29
24.0	1,038	722.2	0.9200	0.01265	44.76	319.36
—	—	662.0	0.9262	0.01250	37.61	326.51
48.0	921	643.3	0.9281	0.01253	35.50	328.62
72.0	867	610.1	0.9316	0.01248	31.92	332.20
96.0	828	583.5	0.9344	0.01244	29.19	334.93
120.0	805	567.9	0.9360	0.01242	27.64	336.48
Well shut in						
0.25	0	728.8	0.9193	0.01266	45.58	—
0.50	0	861.3	0.9062	0.01289	63.65	—
0.75	0	977.4	0.8954	0.01311	81.90	—
1.0	0	1,079.3	0.8863	0.01331	99.71	—
2.0	0	1,378.9	0.8629	0.01399	161.45	—
3.0	0	1,562.4	0.8512	0.01445	205.74	—
4.0	0	1,666.1	0.8457	0.01474	232.78	—
5.0	0	1,730.4	0.8426	0.01492	250.22	—
6.0	0	1,773.2	0.8407	0.01504	262.12	—
24.0	0	1,959.4	0.8341	0.01559	316.31	—
48.0	0	2,011.3	0.8327	0.01575	332.09	—
72.0	0	2,037.3	0.8320	0.01583	340.10	—

The value of *kh* can then be calculated:

$$kh = \frac{1{,}418(450)(555)}{(0.20)(326.51 \times 10^6)}$$

$$kh = 5.42 \text{ md-ft}$$

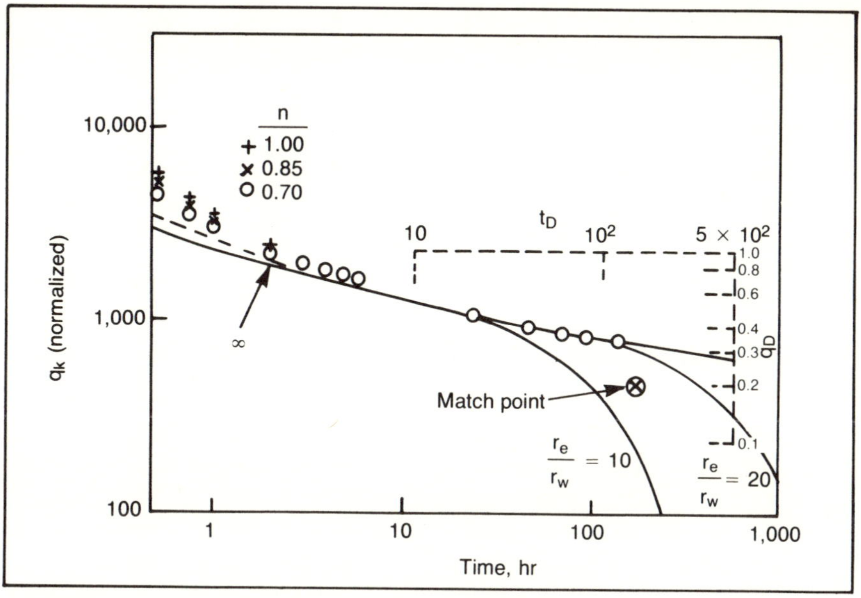

FIGURE 8–3 *Normalized Test Data (See Table 8–5) Imposed on Flow Functions from Figures 8–1A and 8–1B; Real Gas Potential Used*

or:

$$k = 5.42/31 = 0.17 \text{ md}$$

Equation 8–14 can be used unchanged to calculate the equivalent wellbore radius as follows:

$$r_w{}' = \left[\frac{(0.00634)(0.17)(7.29)}{(0.09)(0.01607)(0.000455)(150)}\right]^{0.5} = 8.92 \text{ or } 9 \text{ ft}$$

Note: The use of the real gas potential does not require estimating an average compressibility factor or viscosity. The calculation of the effective wellbore radius requires the use of the viscosity and compressibility of the undisturbed portion of the reservoir. The skin effect, *s*, according to equation 8–11 is:

$$s = -\ell n\left(\frac{r_w'}{r_w}\right) = -\ell n\frac{9}{0.25} = -3.58$$

assuming the wellbore radius as drilled was 0.25 ft.

Working with pressures squared, one previously calculated the *kh* value to be 5.05 md-ft, the permeability to be 0.16 md, and the effective wellbore radius to be 11 ft. With the real gas potential, the values were 5.42, 0.17, and 9 ft, respectively. The agreement is considered good.

CALCULATING THE REAL GAS POTENTIAL, *m(p)*

The real gas potential is calculated starting from a base pressure of 15 psia and at 50, 100, 200, and 300 psia and in 100-psi increments thereafter. The steps are illustrated in table 8–6. The first step is to calculate the compressibility factor, z, and viscosity, μ, for each pressure at the reservoir temperature (in this case 95°F). Compressibility factors are calculated using equations 2–8 to 2–11 and viscosities using equations 2–17 and 2–18. The second step is to calculate $p/\mu z$ for each of the pressures.

By definition in table 8–6, p_m is set at 15 psia and $m(p)$ at 15 psia is set at zero. The $m(p)$ value at 50 psia is:

$$m(50) = 2\left(\frac{4{,}233 + 1{,}267}{2}\right)(50 - 15) + 0 = 0.19250 \times 10^6$$

The $m(p)$ value at 100 psia is:

$$m(100) = 2\left(\frac{8{,}496 + 4{,}233}{2}\right)(100 - 50) + m(50)$$

$$= 0.63645 \times 10^6 + 0.192500 \times 10^6 = 0.82895 \times 10^6$$

The real gas potential at intermediate pressures may be found by interpolation on a plot of $m(p)$ against pressure or calculating the $m(p)$ value at the desired pressure. The inverse of converting values of the real gas potentials to pressures is done most readily by recourse to the plot of $m(p)$ against pressure.

TABLE 8–6.
Example Calculation of the Real Gas Potential, m(p), for the Test Example from Table 7–7,

$\gamma_g = 0.560$, *temperature* = 95°F, *and* M = 16.22

Pressure, p (psia)	Compressibility Factor, z	Viscosity, μ (cp)	p/μz	Δm(p)	m(p)
15	0.9982	0.01186	1,267	0.0	0.0
50	0.9942	0.01188	4,233	0.193 × 10⁶	0.193 × 10⁶
100	0.9883	0.01191	8,496	0.636 × 10⁶	0.829 × 10⁶
200	0.9768	0.01199	17,077	2.557 × 10⁶	3.386 × 10⁶
300	0.9654	0.01209	25,073	4.278 × 10⁶	7.664 × 10⁶
400	0.9543	0.01220	34,357	6.006 × 10⁶	13.670 × 10⁶
500	0.9433	0.01233	42,989	7.735 × 10⁶	21.405 × 10⁶
600	0.9326	0.01247	51,593	9.458 × 10⁶	30.86 × 10⁶
700	0.9222	0.01262	60,142	11.174 × 10⁶	42.04 × 10⁶
800	0.9122	0.01278	68,623	12.877 × 10⁶	54.91 × 10⁶
900	0.9025	0.01296	77,066	14.569 × 10⁶	69.47 × 10⁶
1,000	0.8933	0.01315	85,129	16.220 × 10⁶	85.68 × 10⁶
etc.					

$$m(p) = 2\int_{p_m}^{p} \frac{pdp}{\mu(p)z(p)}$$

PRESSURE BUILDUP BEHAVIOR

Since pressure buildup data were taken (tables 7–7 and 8–5) on the test well after 120 hr of flow, it is desirable to use the buildup data to calculate the permeability and skin effect as a check on the analysis of the drawdown data. The theory of pressure buildup for wells has been given in detail by Matthews and Russell and by Earlougher,[20,21] but the theory given herein will be only that necessary for analysis of the test example given in tables 7–7 and 8–5. The equation for the pressure behavior of a well producing oil above the bubble point after shutin has been given by Matthews and Russell as:

$$p_{ws} = p_i - 162.6\,\frac{q_o \mu B}{kh}\,log\,\frac{(t + \Delta t)}{\Delta t} \tag{8–16}$$

for the case in which the well was producing at a constant rate or a series of discrete constant rates before the well was shut in. Recently, Uraiet and Raghavan have studied the pressure buildup behavior for a well produced against a constant wellbore pressure.[22] They concluded that the same equations can be used for analyzing the buildup from the constant pressure case as has been used for the buildup from the constant rate case, but the production time variable, t, must be defined as the actual production time, and the production rate must be defined as the instantaneous production rate at the time the well was shut in. Equation 8–16 shows that if the pressure, p_{ws}, observed in the well during shutin is plotted against the logarithm, $(t + \Delta t)/\Delta t$, the result should be a straight line with a slope m equal to the coefficient of the logarithmic term from which it is possible to calculate kh:

$$kh = 162.6 \frac{q_o \mu B}{m} \qquad (8\text{–}17)$$

For natural gas engineering, it is convenient to substitute the following for the liquid terms in equations 8–16 and 8–17:

$$q_o \times 5.6146 = 1{,}000\ q_k$$

$$B = \frac{p_b \bar{z} T}{p T_b}$$

$$p = \frac{p_i + p_{ws}}{2}$$

The result is:

$$p_{ws}^2 = p_i^2 - 57{,}920.4 \frac{q_k\ \overline{\mu z}\ T}{kh} \frac{p_b}{T_b} \log \frac{(t + \Delta t)}{\Delta t} \qquad (8\text{–}18)$$

Substituting $p_b = 14.65$ and $T_b = 520$, the result is:

$$p_{ws}^2 = p_i^2 - 1{,}632 \frac{q_k\ \overline{\mu z}\ T}{kh} \log \frac{(t + \Delta t)}{\Delta t} \qquad (8\text{–}19)$$

and:

$$kh = 1{,}632 \frac{q_k \, \overline{\mu z} \, T}{m} \tag{8–20}$$

where the slope, *m*, is expressed in pressure squared per cycle. If the real gas potential, *m*(*p*), is used, equations 8–19 and 8–20 become:

$$m(p)_{ws} = m(p)_i - 1{,}632 \frac{q_k T}{kh} \log \frac{(t + \Delta t)}{\Delta t} \tag{8–21}$$

and:

$$kh = 1{,}632 \frac{q_k T}{m} \tag{8–22}$$

where the slope, *m*, is expressed in the real gas potential, *m*(*p*).

The skin factor, *s*, may be calculated by the following equation given by Earlougher:

$$s = 1.1513 \left[\frac{p_{1hr} - p_{wf}}{m} - \log \left(\frac{k}{\phi \mu \, c_t \, r_w^2} \right) + 3.2275 \right] \tag{8–23}$$

where p_{1hr} is the pressure in the well after 1 hr of shutin time, c_t is the total system compressibility, and r_w is the wellbore radius expressed in feet. In working with natural gas terms, pressure squared and the real gas potential can be substituted for the pressure terms in equation 8–23.

Test Example

Returning to the test example given in table 7–7 and calculating pressures squared and the real gas potential, the results are given in table 8–7 and in figures 8–4 and 8–5. The scales in figures 8–4 and 8–5 were selected to give the appearance of a plot of pressure against time in the buildup, but the changes in no way prevent the use of the build-up equations. Plots such as these are sometimes called Horner plots in deference to the pioneering work of Horner.

TABLE 8–7.
Pressure Build-up Data for Test Example Given in Table 7–7

Shutin Time (hr)	Δt / $(t + \Delta t)$	$\bar{p}_R$ or p_{wf}	$\bar{p}_R^2$ or p_{wf}^2 *(thousands)*	m(p)
0.25	0.002079	728.8	531.1	45.58×10^6
0.50	0.004149	861.3	741.8	63.65×10^6
0.75	0.006211	977.4	955.3	81.90×10^6
1.0	0.008264	1,079.3	1,164.9	99.71×10^6
2.0	0.01639	1,378.9	1,901.4	161.45×10^6
3.0	0.02439	1,562.4	2,441.1	205.74×10^6
4.0	0.03226	1,666.1	2,775.9	232.78×10^6
5.0	0.04000	1,730.4	2,994.3	250.22×10^6
6.0	0.04762	1,773.2	3,144.2	262.12×10^6
24.0	0.1667	1,959.4	3,839.2	316.31×10^6
48.0	0.2857	2,011.3	4,045.3	332.09×10^6
72.0	0.3750	2,037.3	4,150.6	340.10×10^6

Gas produced during test = 5,010 Mcf
Instantaneous rate of flow at end of test = 805 Mcfd
Duration of flow = 120 hr

In attempting to analyze the results presented in figures 8–4 and 8–5, one is confronted with the dilemma of choosing a straight line to use in the analysis. This choice often happens in working with pressure buildup curves for gas wells. The technical literature is replete as to causes of such buildup behavior, and it is no surprise that almost any deviation from the ideal conditions for which the mathematics were derived could be the culprit. In attempting to choose between the straight lines evident in figures 8–4 and 8–5, turn to the work of Earlougher and that of Gringarten, Ramey, and Raghavan.[23,24] The type curve match of the pressure buildup for the test example in table 8–7 on a real gas potential basis from the theoretical work of Gringarten, Ramey, and Raghavan is shown in figure 8–6. The buildup data for 4, 5, and 6 hr can be matched well on the curve for the infinite conductivity vertical fracture, and the data fall to the right of the approximate start of the semi-log straight line portion of the curve. Therefore, one can conclude that the buildup data for 4, 5, 6 hours define the start of the proper portion of the buildup curve to be used in equations 8–20, 8–22, and 8–23 to calculate *kh* and the skin factor, *s*. These have been calculated and are given in table 8–8.

Table 8–8 shows, for comparative purposes, the permeability and the skin factor obtained from the drawdown portion of the well test as well as the results of the pressure squared analysis and the real gas potential analysis. Since the test results given in table 7–7 were taken at random from a number

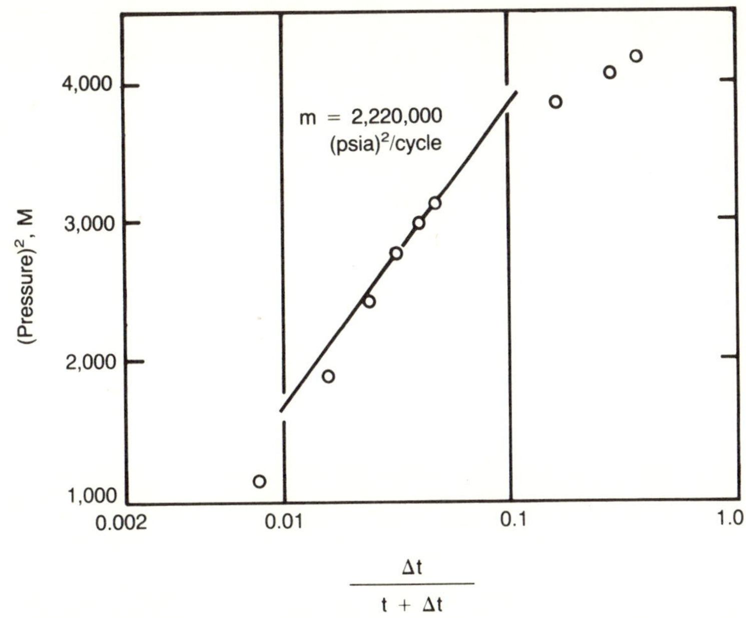

FIGURE 8–4 *Pressure Buildup Using Pressure Squared, Well Test Data from Table 7–7*

of well tests, the agreement between the values for permeability and skin factor given in table 8–8 are believed to be good. The proper straight-line portion of the pressure buildup curve started at about 4 hr of shutin time, which resulted in a line with only three points. More data taken at shutin times of 7, 8, 10, and 12 hr would have been more appropriate for the purpose of the buildup analysis.

However, note that the skin effect as calculated in the foregoing sections includes, in addition to the true skin effects, the effects of non-Darcy flow near the wellbore. To arrive at the true skin effect, it is necessary to determine the apparent skin at several rates of flow and extrapolate to the skin effect indicated at a zero rate of flow. Details of such calculations have been given by Smith.

PRODUCTION FORECASTING

Corollaries to the questions asked early in this chapter about the reserves available for production from the well, are: did the pressure disturbance reach the edge of the reservoir, and was the pressure drop observed at

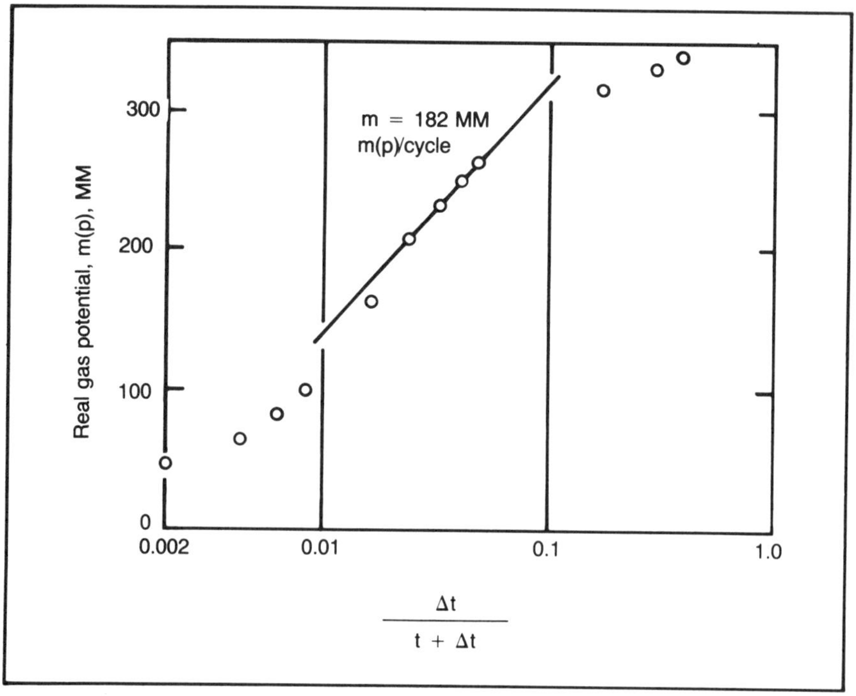

FIGURE 8–5 *Pressure Buildup Using Real Gas Potential, Well Test Data from Tables 7–7 and 8–5*

the well caused by reserve depletion or by the unsteady-state effects? To answer these questions, return to figure 8–3 in which the drawdown data have been imposed on the numerical solutions to the constant back pressure case. The drawdown data follow the curve where the reservoir acts as if it were infinite with respect to the pressure disturbance, and the data points have passed the places where the r_e/r_w lines for values of 10 and 20 diverge from the infinite curve. This means that the pressure disturbance did not reach the reservoir boundary at the total test time of 120 hr and that the external radius of the pressure disturbance was slightly in excess of 20 times the effective wellbore radius. Since the effective wellbore radius was calculated to be 9 ft, using the real gas potential method, the pressure disturbance had receded from the wellbore to a distance of about 9 × 20 = 180 ft or it had covered an area of about 2.34 acres. Thus, the pressure disturbance had not reached a reservoir boundary and had covered a radius of slightly more than 180 ft. Thus far in the test, all pressure changes have been caused by unsteady-state effects and not by an appreciable depletion of the reserves available to the well.

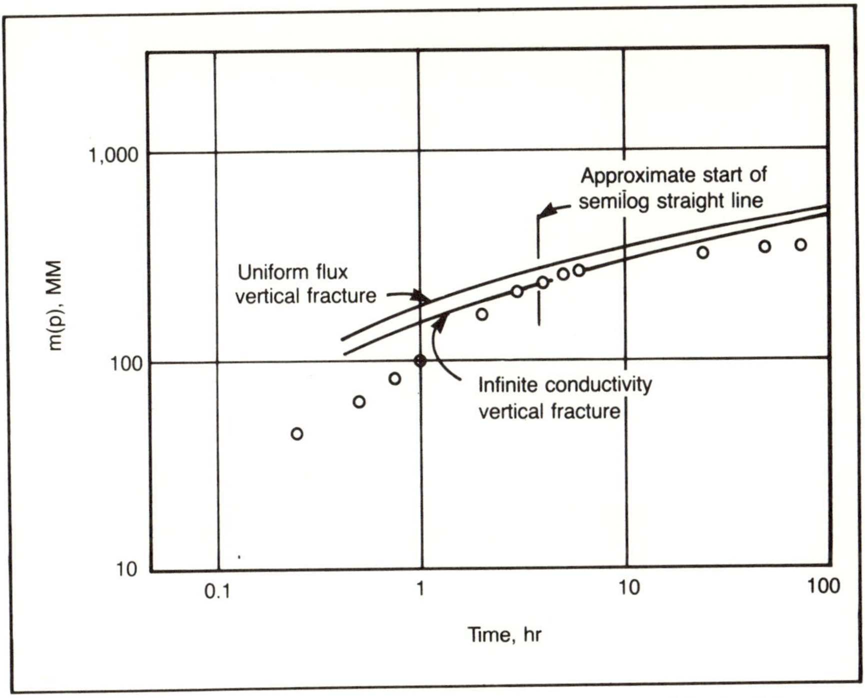

FIGURE 8–6 *Type Curve Match for Infinite Conductivity Vertical Fracture and Pressure Buildup Test Data from Table 8–7*[21,24]

TABLE 8–8.
Summary of Analyses for Permeability and Skin Factor for the Test Example Drawdown and Buildup Results Given in Tables 7–7 and 8–7

	kh, md-ft	Permeability, md	Skin Factor, s
Pressure drawdown			
p^2 analysis	5.05	0.16	−3.78
$m(p)$ analysis	5.42	0.17	−3.58
Pressure buildup			
p^2 analysis	4.14	0.13	−2.90
$m(p)$ analysis	4.01	0.13	−2.87
Thickness, h, = 31 ft			

TIME FOR THE PRESSURE DISTURBANCE TO REACH THE RESERVOIR BOUNDARY

Since the reservoir, insofar as the pressure disturbance is concerned, acts as if it were infinite, one must consult the known geology of the area and the plans for well-spacing in order to calculate the time required for the pressure disturbance to reach the reservoir boundary. If one sees no boundary in his knowledge of the geology and decides that the only boundary foreseen is that imposed by the drainage boundaries of other wells, one can assume a distance of 5,280 ft between wells. The ratio of the external radius to the wellbore radius will be about 2,640/9 = 293, or 300. The dimensionless time, t_D, at which r_e/r_w of 300 diverges from the curve for the infinite system is about 2×10^4 on figure 8–1B, and the dimensionless rate of flow, q_D, will be about 0.183.

Returning to figure 8–3 and recalling that the match point is:

$$q_k = 450 \text{ Mcfd}$$

$$t = 7.29 \text{ days}$$

$$q_D = 0.20$$

$$t_D = 150$$

One can calculate that:

$$t = \frac{7.29}{150} t_D = 0.0486\ t_D$$

Or the time for the pressure disturbance to reach the drainage boundary will be:

$$t = \frac{7.29}{150}(2 \times 10^4) = 972 \text{ days, } 23{,}328 \text{ hr, or } 2.66 \text{ years}$$

At that time, the rate of flow would be:

$$q_k = \frac{450(0.183)}{0.20} = 412 \text{ Mcfd}$$

Returning to table 8–4, the point at which the pressure disturbance reached the drainage boundary has been calculated. After 972 days or 23,328 hr of flow, the rate of flow would be 412 Mcfd and the $(\bar{p}_R^2 - p_{wf}^2)$ value would be 4,031,000. Thus the so-called point of stabilization can be added to figure 7–6, which has been done on figure 8–7.

Production Forecast

The production forecast is made by extending the relationships in figure 8–3 as has been done in figure 8–8, but figure 8–8 returns to the dimensionless rate of flow and time scales. The dimensionless rates of flow and corresponding dimensionless times can be read from the curves in figure 8–8 and then converted from the dimensionless quantities to the real times and rates of flow by using the match point indicated in figure 8–3. This has been done in table 8–9 in which the production rates are instantaneous and are given for the year end.

The production rates of table 8–9 can be averaged over the time periods to give a cumulative production of about 2,800,000 Mcf. Recalling that the original bottom-hole pressure is 2,114.0 psia, the reservoir temperature is 95°F, compressibility factor is 0.8305, the porosity capable of holding gas,

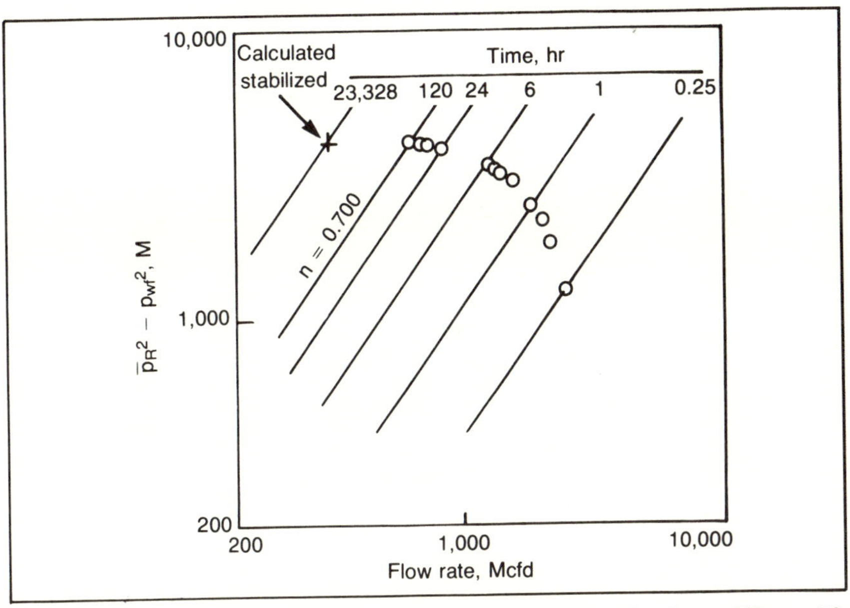

FIGURE 8–7 *Well Test Data from Table 7–7 Plotted as Isochronal Test with Calculated Stabilized Curve Added; Slope is Arbitrary*

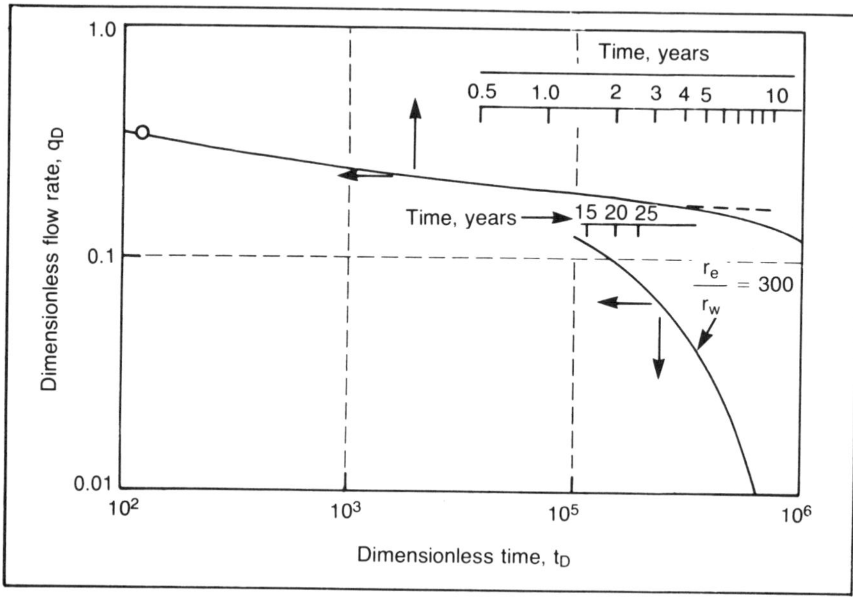

FIGURE 8–8 *Production Forecast for Test Data Given in Table 7–7*

TABLE 8–9.

Production Forecast for the Well from the Test Example in Table 7–7. Production Is Against a Constant Wellbore Pressure, p_{wf}, of 662 psia.

Time (Years)	t_D	q_D	Production *Rate, q_k (Mcfd)*	
0.5	3.75×10^3	0.218	491	
1.0	7.51×10^3	0.201	452	
2.0	1.50×10^4	0.188	423	
2.66	2.00×10^4	0.183	412	(Stabilization)
3.0	2.25×10^4	0.181	407	
4.0	3.00×10^4	0.174	392	
5.0	3.75×10^4	0.167	376	
6.0	4.51×10^4	0.161	362	
7.0	5.26×10^4	0.157	353	
8.0	6.01×10^4	0.150	338	
9.0	6.76×10^4	0.146	329	
10.0	7.51×10^4	0.140	315	
15.0	1.13×10^5	0.118	266	
20.0	1.50×10^5	0.099	223	
25.0	1.88×10^5	0.084	189	

Note: Production rates are given for the year end.

$\phi(1 - S_w)$, is 9%, thickness is 31 ft, and the outer boundary is at 2,640 ft, and the original gas in place from equation 3–5 is:

$$G_k = \frac{1{,}546(0.09)(2{,}114)(31)}{(0.8305)(555)} = 19{,}783 \text{ Mcf/acre}$$

or:

$$G_k = 19{,}783 \times \frac{\pi\,(2{,}640)^2}{43{,}560} = 9{,}940{,}000 \text{ Mcf}$$

The percentage of the gas in place that would be produced in 25 years by the test example well is about 28%. Note that the distance to the reservoir boundary is assumed to be 2,640 ft, or a half mile. Since one of the premises is that the well would be producing against a constant wellbore pressure of 662 psia, the amount of gas that would not be produced under the premise would be:

$$G_k = \frac{1{,}546(0.09)(662)(31)}{(0.9262)(555)} \times \frac{\pi\,(2{,}640)^2}{(43{,}560)} = 2{,}790{,}000 \text{ Mcf}$$

The gas available for production under the constant pressure premise is about 7,150,000 Mcf, and the well will produce about 39% of that quantity in 25 years.

SUMMARY AND CONCLUSIONS

The well of the test example has low productivity, the permeability is about 0.17 md, and the pressure disturbance caused by the testing will never reach an outer reservoir boundary during the test period of 120 hr. In fact, it was calculated that in the 120 hr of testing, the pressure disturbance covered an area of about 2.34 acres. Thus, the question remains unanswered as to whether the reservoir was proved to be commercial by the well test. However, with the ability to make production forecasts, the engineer can estimate the minimum size of the reservoir that is required to meet his standards of economic acceptance. He must then depend upon his knowledge of the geology of the reservoir to assure himself that the reservoir around the well extends to the minimum distance for the well to meet his economic standards.

PROBLEMS

1. Use the isochronal data given below and the remainder of the isochronal data given on the well in table 7–5, a reservoir thickness of 45 ft, an initial reservoir shutin pressure, $\bar{p}_R$, of 441.6 psia, a reservoir temperature of 90°F, a porosity capable of holding gas, $(1 - S_w)$, of 0.12, an outer reservoir boundary of 2,640 ft, and a constant wellbore pressure of 400 psia to calculate:

a. The time required for the pressure disturbance to reach the outer boundary

b. The time and the rate of production schedule for the rate of production to decline to 400 Mcfd and the percent of the original gas in place recovered at that time.

Additional isochronal test data for problem 1:*

Date	*Shutin Pressure, p_R (psia)*	*Duration of Flow (hr)*	*Rate of Flow, q_k (Mcfd)*	*$\bar{p}_R^2 - p_{wf}^2$ (thousands)*
10/11/44	441.6	1.0	1,229	8.62
		9.0	1,202	18.01
		23.5	1,187	23.07
		49.0	1,176	26.71
		70.5	1,171	28.81
		96.5	1,166	30.52
		120.0	1,163	31.56
		144.0	1,161	32.13
		169.0	1,159	32.89
		190.0	1,157	33.54
		214.0	1,156	33.91

2. Recalculate the test example given in the text of chapter 8 from table 7–7 with an outer reservoir boundary of 1,320 ft—in effect reducing the reservoir volume to ¼ that given in the text.

3. In the test example in chapter 8, the effective wellbore radius using the real gas potential was found to be 8.92 ft. If the well were given a fracture treatment to increase the effective wellbore radius to 35 ft, recalculate the time for the pressure disturbance to reach the outer boundary of 2,640 ft and reestimate the production forecast for 25 years.

* All test data were taken from Cullender. (Table 4, copyrighted 1955, Society of Petroleum Engineers of AIME, used by permission.)

REFERENCES

1. A.F. van Everdingen and W. Hurst. "The Application of the Laplace Transform to Flow Problems in Reservoirs." *Trans. AIME*, 186 (1949), p. 305.
2. C.C. Miller, A.B. Dyes, and C.A. Hutchinson Jr. "The Estimation of Permeability and Reservoir Pressure from Bottom Hole Pressure Build-Up Characteristics." *Trans. AIME*, 189 (1950), p. 91.
3. D.R. Horner. "Pressure Build-Up in Wells," *Proc. Third World Pet. Cong.*, ed. E.J. Brill, Leiden, II (1951), p. 503.
4. J.S. Aronofsky and R. Jenkins. "A Simplified Analysis of Unsteady Radial Gas Flow." *Trans. AIME*, 201 (1954), p. 149.
5. R. Jenkins and J.S. Aronofsky. "Unsteady Radial Flow of Gas Through Porous Media." *Jour. Appl. Mech., ASME*, 20 (June 1953).
6. G.H. Bruce, D.W. Peaceman, H.H. Rachford Jr., and J.D. Rice. "Calculation of Unsteady-State Gas Flow Through Porous Media." *Trans. AIME*, 198 (1953), p. 79.
7. R.V. Smith. "Unsteady State Gas Flow into Gas Wells." *Trans. AIME*, 222 (1961), p. 1151.
8. R. Al-Hussainy, H.J. Ramey Jr., and P.B. Crawford. "The Flow of Real Gases Through Porous Media." *Trans. AIME*, 237 (1966), p. 624.
9. Energy Resources Conservation Board. *Gas Well Testing—Theory and Practice*. 4th ed., Calgary, Alberta, Canada, 1979.
10. T.V. Moore, R.J. Schilthuis, and W. Hurst. "The Determination of Permeability from Field Data." *Bull. API*, 211 (May 1933), p. 4.
11. W. Hurst. "Unsteady Flow of Fluids in Oil Reservoirs." *Physics*, 5 (January 1934), p. 20.
12. J. Ferris, D.B. Knowles, R.H. Brown, and R.W. Stallman. *Theory of Aquifer Tests*. Water Supply Paper 1536 E. U.S. Geological Survey (1962), p. 109.
13. K.A. Tsarevich and I.F. Kuranov. *Calculation of the Flow Rates for the Center Well in a Circular Reservoir Under Elastic Conditions, Problems of Reservoir Hydrodynamics*. Part I. Leningrad (1966), pp. 9–34.
14. A.T. Chatas and H. Malekfam. "The Estimation of Aquifer Properties from Reservoir Performance in Water-Drive

Fields.'' SPE 2970. Presented at 45th Annual Fall SPE Meeting, Houston, Texas, October 4–7, 1970.

15. M.J. Fetkovich. ''Decline Curve Analysis Using Type Curves.'' *J. Pet. Tech; Trans. SPE*, 269 (June 1980), p. 1065.
16. M.J. Fetkovich and T.S. Thrasher. ''Constant Well Pressure Testing and Analysis in Low Permeability Reservoirs.'' SPE 7928.
17. R.G. Agarwal, R.D. Carter, and C.B. Pollock. ''Evaluation and Performance Prediction of Low-Permeability Gas Wells Stimulated by Massive Hydraulic Fracturing.'' *J. Pet. Tech., Trans. AIME*, 267 (March 1979), 363.
18. A.F. van Everdingen. ''The Skin Effect and Its Influence on the Productive Capacity of a Well.'' *Trans. AIME*, 198 (1953), p. 171.
19. A.G. Winestock and G.P. Colpitts. ''Advances in Estimating Gas Well Deliverability.'' *J. Can. Pet. Tech.*, 4 (1965), p. 111.
20. C.S. Matthews and D.G. Russell. *Pressure Buildup and Flow Tests in Wells.* Monograph Series, vol. 1. SPE of AIME, Dallas (1967).
21. R.C. Earlougher Jr. *Advances in Well Test Analyses*, Monograph Series, vol. 5. SPE of AIME, Dallas (1977).
22. A.A. Uraiet and R. Raghavan. ''Pressure Buildup Analysis for a Well Produced at a Constant Bottomhole Pressure.'' *J. Pet. Tech., Trans. SPE*, 269 (October 1980), p. 1813.
23. A.C. Gringarten, H.J. Ramey Jr., and R. Raghavan. ''Pressure Analysis for Fractued Wells.'' SPE 4051, presented at the SPE-AIME 47th Annual Fall Meeting, San Antonio, Texas, October 8–11, 1972.
24. A.C. Gringarten, H.J. Ramey Jr., and R. Raghavan. ''Unsteady-State Pressure Distributions Created by a Well With a Single Infinite-Conductivity Vertical Fracture.'' *Soc. Pet. Eng. J. Trans. AIME*, 257 (August 1974), p. 347.

9 Production Forecasting for Gas Wells

In the previous chapter, the unsteady-state theory of gas flow was used to predict the future production from a gas well when the well was produced at capacity against a constant back pressure or pipeline pressure. To be able to make the prediction for that particular well, the following were needed:

1. An information test of at least 120-hr duration (for this particular well)
2. Independent estimates of the porosity, thickness, and water saturation in the reservoir
3. An estimate of the outer radius of drainage since the test on the well did not last long enough for the pressure disturbance to reach a boundary

Note that a small change in the 662-psia back pressure would make little change in the results (see figure 7–3). Nevertheless, the important feature of the example in chapter 8 was that the well was operating at capacity against a constant back pressure. This is a common mode of operation for many marginal wells or wells with low production capacities. It also applies to many wells located near a large pipeline that serves many wells. It is true that the constant back pressure can be changed with the use of compressors, but there may be a time when further change in the back pressure is no longer feasible from an economic standpoint.

OPERATING CONDITIONS FOR PRODUCTION FORECASTING

In considering the operation of gas wells in general, three principal modes of operation can be identified:

1. The well has excess production capacity and production is restricted to a constant daily (or monthly) quantity of gas, or in other words, production is restricted to the market demand.
2. Production is allocated to various wells, usually located in a proration unit, to a fraction of the open-flow potential or according to some definition of deliverability, or the wells are allowed to produce with a wellhead pressure that is at a constant fraction of the current shutin wellhead pressure.
3. The well produces against a constant wellhead pressure.

Other modes of operation can be devised, but the author believes that most gas wells are operated in such a manner as to approximate one of the modes mentioned in the foregoing list. In order to make a production forecast for a well, it is necessary to idealize the operations to some extent. In the production forecast example of the previous chapter, there is no way in reality that a well could be turned on and allowed to produce continuously for 25 years against a constant back pressure as indicated in the production forecast.

The daily average production or take from gas wells is usually set by two procedures or a combination of the two. These are the contract quantity and the allowable assigned to the well by a regulatory agency. The contract quantity is the amount of gas usually specified in a legal document that the producer has agreed to take from the well or a group of wells. It usually can be stated as a daily average quantity, but in reality the amount of gas that can be taken by the purchaser may vary from day to day, depending upon the purchaser's market demand. If there are competing producers or competing purchasers in a common reservoir, there usually exists a set of rules that specifies the amount of gas to be produced by the well over a period of time, usually a month.

If the well is capable of delivering more gas into the pipeline than the purchaser requires, production from the well must be restrained and the well is said to have excess productive capacity.

Regarding the second mode of operation, recall that the allowables for many gas wells are limited to 25% of the current open-flow potential. Also, many wells are operated in such a manner that the flowing pressure at the wellhead is an almost-constant fraction of the shutin pressure. In order to understand why this should occur, the reader is referred to figure 7–3 in which the flowing pressure for a gas well is shown as a fraction of the shutin pressure and the resulting rate of flow is shown as a fraction of maximum producing capacity. If one assumes that the well is being operated so the flowing pressure is 50% of the shutin pressure, the well is producing at 75–86% of its productive capacity. Further reduction in the flowing pressure results in little increase in the rate of production. If a group of wells is being produced into a common compressor, the collective rate of production soon reaches a state of balance with the compressor operation so that a further increase in compressor capacity is not justified by the increase in production.

The third mode is operating against a constant back pressure. This type of operation usually occurs where a gas well or a small group of gas wells is near a large pipeline carrying a large quantity of gas. In this instance, the question will be how much gas can be produced into the pipeline before compression is justified.

In forecasting production from a well, the engineer must decide how the well will be operated. The first task is to decide how the average daily take from the well will be determined. If the well is in a prorated field, the decision may have already been made. If so, the average daily allowable should be determined and compared to the productive capacity of the gas well against the expected pipeline pressure at the wellhead. If the well has excess productive capacity, the problem becomes one of determining how much gas can be produced until the productive capacity of the well is equal to the allowable or average daily take from the well. If the operator of the well and the purchaser of the gas have reached agreement as to the daily average contract quantity and there are no other restrictions imposed by regulatory agencies, the problem remains the same.

In developing procedures to make production forecasts for the three listed modes of operation, one will need the following information:

1. A stabilized back-pressure or stabilized isochronal performance for the well
2. An independent estimate of the quantity of gas in place available for production from the well
3. A determination of the market demand for gas from the well

WELLS WITH EXCESS PRODUCTIVE CAPACITY

A well with excess productive capacity can deliver more gas than the purchaser is willing to take. In other words, the contract quantity is less than the capacity of the well to deliver gas into the sales pipeline against the pressure in the pipeline. If the back-pressure equation is known for the wellhead performance of the well when the pressure disturbance reaches the reservoir boundary (the well is said to be stabilized when the disturbance reaches the boundary), one can calculate the minimum shutin pressure at the wellhead at which the well can produce a given quantity of gas into the pipeline operating at a given pressure. If the performance curve at stabilization is:

$$q_k = C(\bar{p}_{ts}{}^{z} - p_{tf}{}^2)^n \quad (9\text{–}1)$$

The pipeline pressure is p_{tfp} and the minimum shutin pressure against which the well can produce the market demand, q_{kmd}, is:

$$p_{ts}\,(minimum) = (\Delta\bar{p}^2(1{,}000) + p_{tfp}{}^2)^{1/2} \quad (9\text{–}2)$$

where Δp^2 is the difference in pressure squared in thousands for the market demand rate of flow.

In order to illustrate the calculations, consider a hypothetical well that has a stabilized wellhead performance curve with pressure squared in thousands as follows:

$$q_k = 12.0\,(\bar{p}_{ts}{}^2 - p_{tf}{}^2)^{0.70} \quad (9\text{–}3)$$

The contract quantity is 1,250 Mcfd, the wellhead shutin pressure is 1,500 psia, and the delivery pressure or pipeline pressure is 1,000 psia. First, calculate the shutin pressure for the well when the well can no longer produce its contract quantity of gas into the pipeline. If one substitutes the present shutin pressure and the back pressure or pipeline pressure into equation 9–3, the well is capable of producing 1,766 Mcfd into the pipeline:

$$q_k = 12.0\left(\frac{1{,}500^2 - 1{,}000^2}{1{,}000}\right)^{0.70} = 1{,}766 \text{ Mcfd}$$

(**Note:** The custom of expressing pressure squared in thousands has been followed.) Thus, the well is capable of producing more than its contract quantity of 1,250 Mcfd by a margin of 41%. The difference in pressure squared in thousands required for a production rate of 1,250 Mcfd is:

$$1{,}250 = 12.0(\Delta p^2)^{0.7}$$

$$\Delta p^2 = (1{,}250/12.0)^{1/0.7} = 762.9 \text{ M}$$

Substituting this into equation 9–2:

$$p_{ts}\,(minimum) = (762{,}900 + 1{,}000^2)^{1/2} = 1{,}327.7 \text{ psia}$$

If one makes the rough approximation that the material balance can be based on shutin wellhead pressures, about (1–1,327.7/1,500) × 100, or 11%, of the gas in place can be produced at a constant rate of 1,250 Mcfd. A more realistic estimate of the gas to be produced would be to calculate the shutin pressures in the reservoir corresponding to the wellhead pressures of 1,500 and 1,327.7 psia, estimate the corresponding compressibility factors, and calculate the gas to be produced by:

$$G_p = \frac{p_1/z_1 - p_2/z_2}{p_1/z_1} G_1 \qquad (9\text{–}4)$$

where G_p is the gas produced to reduce the pressure from p_1 to p_2 in the reservoir and G_1 is the gas in place when the reservoir pressure was p_1.

The time interval that the well would produce at the constant rate of 1,250 Mcfd would be found by dividing the constant producing rate into the quantity of gas produced according to equation 9–5:

$$t = G_p/q \qquad (9\text{–}5)$$

For example, if the gas in place available for production from the well is 2.5 billion cu ft, and if the estimate that 11% of the gas in place is accurate enough for the purpose at hand, the time of production at 1,250 Mcfd would be:

$$t = \frac{(2.5 \times 10^9)(0.11)}{1{,}250 \times 10^3} = 220 \text{ days}$$

The well would produce (1,250)(220) = 275,000 Mcf of gas during the interval.

WELLS WITH PRODUCTION RATE LIMITED TO A FRACTION OF THE OPEN-FLOW POTENTIAL

It can be shown that production at a constant fraction of the open-flow capacity or potential is identical mathematically to production against a constant fraction of the current shutin pressure. From equation 7–1:

$$q = C(\bar{p}_R^{\,2} - p_{wf}^{\,2})^n \qquad (7\text{–}1)$$

The open-flow capacity is:

$$q_{of} = C(\bar{p}_R{}^2)^n \tag{9–6}$$

Dividing:

$$\frac{q}{q_{of}} = \left[1 - \left(\frac{p_{wf}}{\bar{p}_R}\right)^2\right]^n = constant \tag{9–7}$$

If (q/q_{of}) is a constant, $p_{wf}/\bar{p}_R$ is also a constant. This same relationship was developed in equation 7–9 and illustrated in figure 7–3 for the wellhead curves. It was repeated here for emphasis.

If $p_{wf}/\bar{p}_R = W$ is set as a constant, equation 7–1 can be rewritten as:

$$q = C\,[\bar{p}_R{}^2\,(1 - W^2)]^n$$

Since:

$$q = -\frac{dG}{dt}$$

Then:

$$-\frac{dG}{dt} = C[\bar{p}_R{}^2\,(1 - W^2)]^n \tag{9–6}$$

At this point, one must choose the relationship to use between G, the gas in place,and $\bar{p}_R$, the average reservoir pressure in the gas reserve available for production from the well. Making the assumption that the compressibility factor, z, is 1.00:

$$C = \frac{q_i}{[\bar{p}_{Ri}{}^2\,(1 - W^2)]^n}$$

This allows the subscript i to denote initial conditions and integrates between initial conditions when $t = 0$ and $\bar{p}_R$ is at t. The result is:

$$\frac{q}{q_i} = \frac{1}{\left[\frac{(2n - 1)\, q_i t}{G_i} + 1\right]^{\frac{2n}{2n - 1}}} \text{ and } n \neq 0.5 \qquad (9\text{–}9)$$

This is the same as equation 30, published by Fetkovich, except for an error in the sign in the denominator of the exponent in the published equation.[1]

If n, the exponent in the back-pressure equation, is 0.5, the equation relating q to q_i becomes:

$$\frac{q}{q_i} = \frac{1}{exp\left(\frac{q_i t}{G_i}\right)} \text{ and } n = 0.5 \qquad (9\text{–}10)$$

Equations 9–9 and 9–10 can be integrated against time by:

$$G_p = \int_o^t q\, dt \qquad (9\text{–}11)$$

Equation 9–9 becomes:

$$\frac{G_p}{G_i} = 1 - \left[\frac{(2n - 1)\, q_i t}{G_i} + 1\right]^{\frac{1}{1 - 2n}} \text{ and } n \neq 0.5 \qquad (9\text{–}12)$$

Equation 9–10 becomes:

$$\frac{G_p}{G_i} = 1 - exp\left(-\frac{q_i t}{G_i}\right) \textit{ and } n = 0.5 \qquad (9\text{–}13)$$

Equations 9–10, 9–12, and 9–13 have been published by Fetkovich.

Note: The equations relating producing rate, q, to time do so in terms of dimensionless rate, q/q_i, and dimensionless time, $(q_i t)/G_i$. Therefore, solutions for the equations can be given conveniently as in table 9–1.

The solutions have also been shown in figures 9–1, 9–2, and 9–3 for n values of 0.5, 0.7 and 1.0. Two features are worthy of note. First, the curves

TABLE 9–1.
Dimensionless Rate of Flow, q/q_i, as a Function of Dimensionless Time, $(q_i t)/G_i$, for Wells Flowing at a Constant Fraction of Open-Flow Capacity or against a Constant Fraction of Current Shutin Pressure

Values of $q_i t/G_i$

	n					
q/q_i	1.000	0.900	0.800	0.700	0.600	0.500
1.00	0.000	0.000	0.000	0.000	0.000	0.000
0.95	0.026	0.029	0.032	0.037	0.043	0.051
0.90	0.054	0.060	0.067	0.076	0.089	0.105
0.80	0.118	0.130	0.145	0.165	0.189	0.223
0.70	0.195	0.215	0.239	0.268	0.306	0.357
0.60	0.291	0.319	0.352	0.393	0.444	0.511
0.50	0.414	0.451	0.495	0.548	0.612	0.693
0.40	0.581	0.628	0.683	0.748	0.825	0.916
0.35	0.690	0.743	0.804	0.874	0.956	1.050
0.30	0.826	0.885	0.951	1.026	1.111	1.204
0.25	1.000	1.065	1.136	1.215	1.300	1.386
0.20	1.236	1.306	1.381	1.460	1.538	1.609
0.15	1.582	1.655	1.728	1.799	1.859	1.897
0.10	2.162	2.228	2.286	2.327	2.339	2.303
0.07	2.780	2.826	2.851	2.845	2.788	2.659
0.05	3.472	3.483	3.459	3.384	3.238	2.996
0.03	4.774	4.689	4.541	4.308	3.970	3.507
0.015	7.165	6.832	6.384	5.800	5.068	4.200
0.010	9.000	8.428	7.706	6.819	5.772	4.605

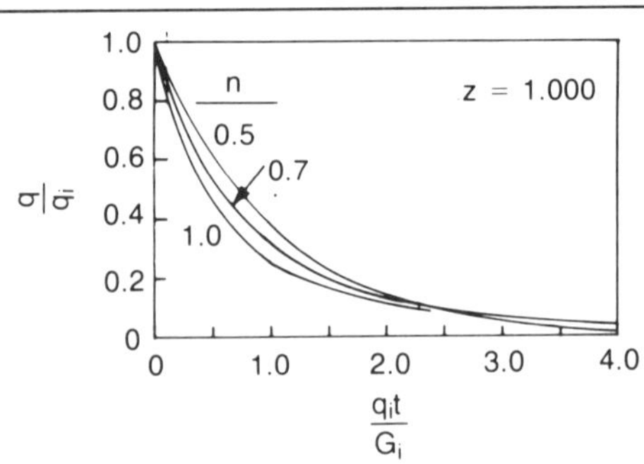

FIGURE 9–1 *Producing Rates for Wells Flowing at Constant Fraction of Open Flow Capacity or Constant Fraction of Current Shutin Pressure*

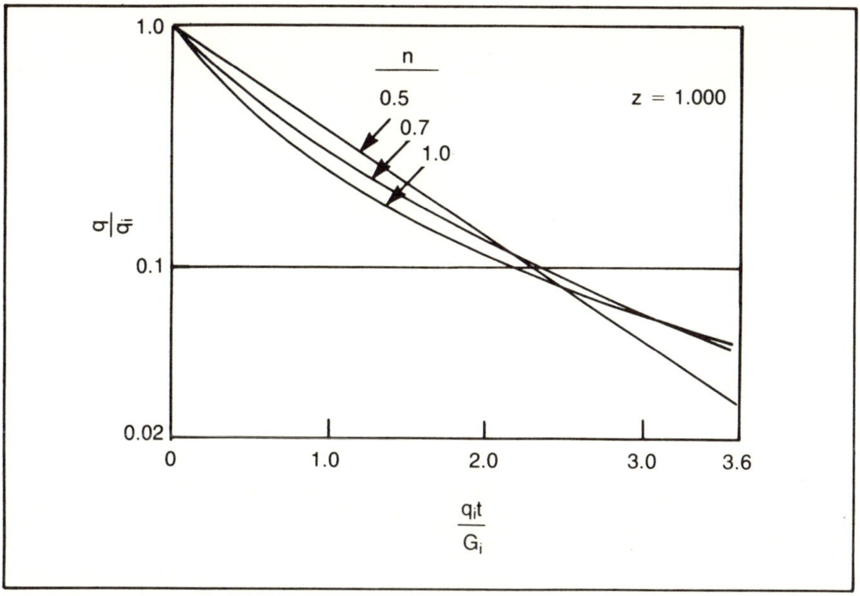

FIGURE 9–2 *Producing Rates for Wells Flowing at Constant Fraction of Open Flow Capacity or Constant Fraction of Current Shutin Pressure*

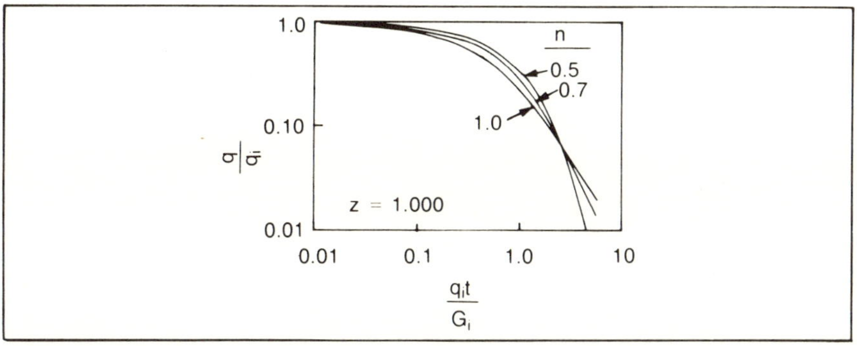

FIGURE 9–3 *Producing Rates for Wells Flowing at Constant Fraction of Open Flow Capacity or Constant Fraction of Current Shutin Pressure*

cross at the dimensionless time, $q_i t/G_i$, values of about 2.0 and continue to spread beyond that value. Second and most important, the dimensionless rate of flow, q/q_i, is a straight-line function of the time function, $q_i t/G_i$, for an n value of 0.5 on semilogarithmic coordinates as shown on figure 9–2.

It has been common practice to extrapolate declining production rates from a gas well on semilogarithmic coordinates to estimate future production rates. Since the curves for $n > 0.5$ on figure 9–2 do not depart much from the straight line for $n = 0.5$, there is support for extrapolating the trend of production rates for gas wells as straight lines on semilogarithmic coordinates.

Numerical Example

In the previous example, it was calculated that a hypothetical gas well would produce 1,250 Mcfd for 220 days against a constant pressure of 1,000 psia before it reached the breaking point. There, if the back pressure were not lowered, the rate of flow would decrease; if the rate of flow were held constant, it would be necessary to lower the back pressure continuously. At the breaking point, the gas remaining in place was 2,500,000,000 cu ft − 275,000,000 cu ft = 2,225,000,000 cu ft or 2.225×10^9 cu ft.

Recalling that the exponent, n, was 0.70, equation 9–9 can be used to calculate the rates of flow at various times, providing the ratio of the flowing pressure to the current shutin pressure remains constant throughout the time period for which the calculations are made. At a flow rate of 1,250 Mcfd, the difference in pressure squared is 762.9 thousands, and the shutin pressure is calculated to be 1,327.7 psia. Thus, at the breaking point the ratio of the flowing pressure to the current shutin pressure is:

$$1{,}000/1{,}327.7 = 0.7532$$

The value of q_i for use in equation 9–9 is 1,250 Mcfd and G_i is 2.225×10^9 cu ft ($z = 1.00$ was assumed). Equation 9–9 reduces to:

$$\frac{q(365)}{(1.25 \times 10^6)(365)} = \frac{1}{\left[\dfrac{(0.4)(1.25 \times 10^6)(365)t}{2.225 \times 10^9} + 1\right]^{3.5}} \qquad (9\text{–}14)$$

At this point, one must be careful with the time units. If time units in years are used, multiply the daily rate in equation 9–9 by 365. (**Note:** The 365 cancels out of the left side of the equation.) Equation 9–14 then reduces to:

$$q = \frac{1.25 + 10^6}{(0.08202t + 1)^{3.5}} \qquad (9\text{–}15)$$

where the time, t, is expressed in years. The results are given in the second column of table 9–2 for a period of 8 yr. The total gas produced to a given time can be calculated using equation 9–12:

TABLE 9–2.
Production Forecast for a Well Producing Against a Back Pressure that is a Constant Fraction of its Current Shutin Pressure

q_i = 1,250 Mcfd   *Initial Shutin Pressure* = 1,327.7 psia
$G_i\ q_i$ = 2.225 × 10⁹ cu ft   *Initial Flowing Pressure* = 1,000.0 psia
n = 0.70 z = 1.00

t, Time (years)	q_k At Year End (Mcfd)	G_p, Gas Production (Mcf)	q_k, Average Production Rate (Mcfd)	G_p, Cumulative Gas Production (Mcf)
0	1,250	0	0	0
1	949	397,986	1,090	397,986
2	735	305,029	836	703,015
3	579	238,228	653	941,244
4	463	189,121	518	1,130,365
5	375	152,300	417	1,282,665
6	308	124,208	340	1,406,873
7	255	102,442	281	1,509,315
8	214	85,345	234	1,594,661

$$\frac{G_p}{2.225 \times 10^9} = 1 - \left[\frac{(0.4)(1.25 \times 10^6)(365)t}{2.225 \times 10^9} + 1\right]^{-2.5} \qquad (9\text{–}16)$$

or:

$$G_p = 2.225 \times 10^9(1 - [0.08202t + 1]^{-2.5}) \qquad (9\text{–}17)$$

Again, t is expressed in years, and the results are given in the last column of table 9–2.

Since the foregoing numerical example is a continuation of the previous example in which the well was produced against a constant back pressure of 1,000 psia, the time scale has been combined on semilogarithmic coordinates for figure 9–4. The portion of figure 9–4 labeled *a* shows production when the well had excess capacity, and *b* shows production after the breaking point when the well was produced against a pressure that was a constant fraction of the current shutin pressure.

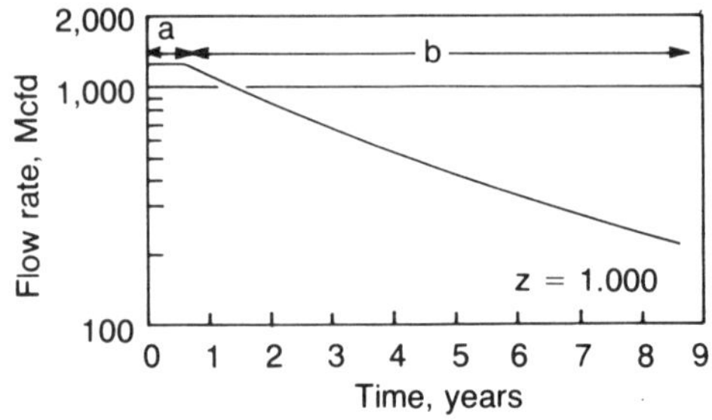

FIGURE 9–4 *Producing Rate for a Gas Well (a) Producing a Constant Rate Against a Constant Back Pressure Until the Breaking Point Thereafter (b) Against a Back Pressure That Is a Constant Fraction of the Current Shutin Pressure*

PRODUCTION RATES FOR WELLS FLOWING AGAINST A CONSTANT BACK PRESSURE

Starting with the back pressure equation 7–1:

$$q = C(\bar{p}_R^{\,2} - p_{wf}^{\,2})^n \qquad (7\text{–}1)$$

and setting $x = \bar{p}_R/p_{wf}$. Since p_{wf} is a constant, one can derive the following equation by setting the compressibility factor equal to 1.00:

$$\frac{dx}{(x^2 - 1)^n} = -\frac{q_i\, x_i\, dt}{G_i(x_i^{\,2} - 1)^n} \qquad (9\text{–}18)$$

at $t = 0$, $x = x_i$, and at t, $x = x$.

If n is 1.0, equation 9–18 can be integrated between limits to give:

$$\frac{(x_i - 1)(x + 1)}{(x_i + 1)(x - 1)} = exp\left[\frac{x_i}{(x_i^{\,2} - 1)}\right]\left[\frac{q_i t}{G_i}\right] \qquad (9\text{–}19)$$

If n is 0.5, equation 9–18, on integration, becomes:

$$\frac{x_i + (x_i^{\,2} - 1)^{0.5}}{x + (x^2 - 1)^{0.5}} = exp\left[\frac{x_i}{(x_i^{\,2} - 1)^{0.5}}\right]\left[\frac{q_i t}{G_i}\right] \qquad (9\text{–}20)$$

However, if n is between 0.5 and 1.0, one cannot find an analytical solution and must resort to numerical integration. Equation 9–18 then becomes:

$$\int_{x_i}^{x} \frac{dx}{(x^2 - 1)^n} = \left[\frac{x_i}{(x_i^2 - 1)^n}\right]\left[\frac{q_i t}{G_i}\right] \qquad (9\text{–}21)$$

Or it can be written as:

$$A_x - A_{xi} = \left[\frac{x_i}{(x_i^2 - 1)^n}\right]\left[\frac{q_i t}{G_i}\right] \qquad (9\text{–}22)$$

Arbitrarily, A_x can be defined as:

$$A_x = \int_{10}^{x} \frac{dx}{(x^2 - 1)^n} \qquad (9\text{–}23)$$

Values of A_x have been calculated at convenient intervals of x and are given in table 9–3. Rates of flow corresponding to x can be calculated by the following equation:

$$q = q_i \left[\frac{x^2 - 1}{x_i^2 - 1}\right]^n \qquad (9\text{–}24)$$

At this point, note the relationship between equation 9–21 for wells producing against a constant back pressure and equation 9–9 for wells producing against a back pressure that is a constant fraction of the current shutin pressure. If one considers x^2 large with respect to 1 in equation 9–21, equation 9–9 can be derived from equation 9–21. In other words, the limit of equation 9–21 is equation 9–9 as x^2 approaches a large number. Another way of saying the same thing is that equation 9–21 and equation 9–9 become equivalent as the constant back pressure approaches zero.

TABLE 9–3.

Values of $A_x = \int_{10}^{x} \frac{dx}{(x^2 - 1)^n}$ *for Wells Producing Against a Constant Back Pressure*

x	n 0.50	0.60	0.70	0.80	0.90	1.00
1.004	2.9038	2.6152	2.4864	2.5032	2.6695	3.0080
1.005	2.8933	2.5983	2.4593	2.4598	2.6001	2.8966
1.006	2.8837	2.5834	2.4358	2.4230	2.5422	2.8057
1.007	2.8750	2.5698	2.4149	2.3907	2.4924	2.7289
1.008	2.8668	2.5574	2.3960	2.3620	2.4487	2.6624
1.009	2.8592	2.5459	2.3788	2.3360	2.4097	2.6037
1.01	2.8519	2.5352	2.3628	2.3123	2.3744	2.5513
1.02	2.7936	2.4520	2.2442	2.1433	2.1333	2.2072
1.03	2.7489	2.3917	2.1629	2.0334	1.9850	2.0070
1.04	2.7113	2.3428	2.0991	1.9502	1.8765	1.8656
1.05	2.6783	2.3008	2.0458	1.8826	1.7906	1.7565
1.06	2.6485	2.2638	1.9997	1.8253	1.7193	1.6677
1.07	2.6212	2.2304	1.9589	1.7753	1.6583	1.5931
1.08	2.5958	2.1999	1.9221	1.7310	1.6048	1.5287
1.09	2.5721	2.1716	1.8885	1.6910	1.5573	1.4722
1.10	2.5497	2.1452	1.8575	1.6546	1.5145	1.4219
1.11	2.5284	2.1205	1.8287	1.6212	1.4756	1.3766
1.12	2.5081	2.0971	1.8018	1.5902	1.4399	1.3355
1.13	2.4887	2.0750	1.7765	1.5613	1.4069	1.2978
1.14	2.4701	2.0539	1.7526	1.5342	1.3763	1.2631
1.15	2.4521	2.0337	1.7300	1.5088	1.3476	1.2310
1.16	2.4380	2.0144	1.7084	1.4847	1.3208	1.2010
1.17	2.4181	1.9959	1.6879	1.4619	1.2955	1.1730
1.18	2.4019	1.9780	1.6682	1.4403	1.2717	1.1467
1.19	2.3861	1.9608	1.6493	1.4196	1.2491	1.1220
1.20	2.3709	1.9441	1.6312	1.3999	1.2276	1.0986
1.21	2.3560	1.9280	1.6138	1.3810	1.2072	1.0765
1.22	2.3415	1.9124	1.5970	1.3629	1.1877	1.0555
1.23	2.3274	1.8973	1.5807	1.3455	1.1690	1.0355
1.24	2.3136	1.8826	1.5650	1.3288	1.1512	1.0165
1.25	2.3001	1.8682	1.5498	1.3126	1.1341	0.9983
1.26	2.2869	1.8543	1.5351	1.2971	1.1176	0.9809
1.27	2.2740	1.8407	1.5208	1.2820	1.1018	0.9642
1.28	2.2613	1.8275	1.5069	1.2675	1.0865	0.9482
1.29	2.2489	1.8145	1.4934	1.2534	1.0718	0.9329

Table 9–3, continued

	n					
x	*0.50*	*0.60*	*0.70*	*0.80*	*0.90*	*1.00*
1.30	2.2368	1.8019	1.4803	1.2397	1.0576	0.9181
1.35	2.1792	1.7427	1.4194	1.1770	0.9931	0.8518
1.40	2.1262	1.6890	1.3651	1.1221	0.9376	0.7955
1.45	2.0769	1.6399	1.3161	1.0733	0.8888	0.7470
1.50	2.0308	1.5945	1.2714	1.0293	0.8456	0.7044
1.55	1.9874	1.5523	1.2304	0.9894	0.8067	0.6666
1.60	1.9463	1.5127	1.1924	0.9528	0.7716	0.6328
1.65	1.9072	1.4756	1.1570	0.9192	0.7396	0.6023
1.70	1.8700	1.4405	1.1239	0.8880	0.7102	0.5746
1.75	1.8344	1.4073	1.0929	0.8590	0.6831	0.5493
1.80	1.8003	1.3757	1.0636	0.8319	0.6580	0.5260
1.85	1.7676	1.3456	1.0359	0.8064	0.6346	0.5046
1.90	1.7360	1.3168	1.0097	0.7825	0.6128	0.4847
1.95	1.7056	1.2893	0.9848	0.7600	0.5924	0.4662
2.00	1.6763	1.2629	0.9611	0.7386	0.5732	0.4490
2.10	1.6204	1.2131	0.9169	0.6992	0.5381	0.4177
2.20	1.5678	1.1669	0.8761	0.6635	0.5067	0.3901
2.30	1.5182	1.1238	0.8386	0.6309	0.4783	0.3654
2.40	1.4711	1.0833	0.8038	0.6010	0.4526	0.3433
2.50	1.4264	1.0453	0.7714	0.5734	0.4291	0.3233
2.60	1.3838	1.0093	0.7411	0.5478	0.4075	0.3051
2.70	1.3430	0.9752	0.7126	0.5260	0.3877	0.2885
2.80	1.3040	0.9429	0.6858	0.5018	0.3693	0.2733
2.90	1.2665	0.9121	0.6605	0.4810	0.3522	0.2592
3.00	1.2305	0.8827	0.6366	0.4615	0.3362	0.2462
3.25	1.1460	0.8147	0.5818	0.4174	0.3007	0.2177
3.50	1.0684	0.7533	0.5332	0.3789	0.2703	0.1936
3.75	0.9966	0.6973	0.4896	0.3449	0.2438	0.1729
4.00	0.9298	0.6460	0.4502	0.3147	0.2206	0.1551
4.25	0.8673	0.5987	0.4143	0.2875	0.2000	0.1395
4.50	0.8086	0.5547	0.3814	0.2629	0.1815	0.1257
4.75	0.7532	0.5138	0.3511	0.2404	0.1650	0.1134
5.00	0.7008	0.4754	0.3231	0.2199	0.1499	0.1024
5.25	0.6510	0.4394	0.2970	0.2010	0.1363	0.0925
5.50	0.6037	0.4055	0.2727	0.1836	0.1237	0.0835
5.75	0.5585	0.3734	0.2499	0.1674	0.1123	0.0754
6.00	0.5153	0.3430	0.2285	0.1523	0.1017	0.0679
6.25	0.4740	0.3141	0.2083	0.1383	0.0918	0.0611
6.50	0.4342	0.2866	0.1893	0.1251	0.0827	0.0547
6.75	0.3961	0.2604	0.1713	0.1127	0.0742	0.0489
7.00	0.3593	0.2353	0.1542	0.1011	0.0663	0.0435

Table 9–3, continued

x	n 0.50	0.60	0.70	0.80	0.90	1.00
7.25	0.3239	0.2114	0.1380	0.0901	0.0589	0.0385
7.50	0.2896	0.1884	0.1226	0.0797	0.0519	0.0338
7.75	0.2566	0.1663	0.1078	0.0699	0.0453	0.0294
8.00	0.2246	0.1451	0.0938	0.0606	0.0392	0.0253
8.25	0.1936	0.1247	0.0803	0.0517	0.0333	0.0215
8.50	0.1635	0.1050	0.0675	0.0433	0.0278	0.0179
8.75	0.1343	0.0860	0.0551	0.0352	0.0226	0.0145
9.00	0.1060	0.0676	0.0432	0.0276	0.0176	0.0112
9.25	0.0784	0.0499	0.0318	0.0202	0.0129	0.0082
9.50	0.0516	0.0327	0.0208	0.0132	0.0084	0.0053
9.75	0.0254	0.0161	0.0102	0.0065	0.0041	0.0026
10.00	0.0000	0.0000	0.0000	0.0000	0.0000	0.0000

Numerical Example

Previously, an example of a hypothetical gas well that produced 1,250 Mcfd against a constant pressure of 1,000 psia for a total of 220 days was used. At the breaking point for this well, the current shutin pressure was calculated to be 1,327.7 psia, and the gas in place (in this case G_i) was 2.225 × 10⁹ cu ft. Now calculate a production forecast for the well if it continues to produce against the constant 1,000 psia back pressure. At time zero, $x_i = 1{,}327.7 \div 1{,}000 = 1.3277$, and A_{xi}, interpolated from table 9–3, is 1.4466 for an n value of 0.70. Equation 9–22 then becomes:

$$A_x - 1.4466 = \left[\frac{1.3277}{(1.3277^2 - 1)^{0.7}}\right]\left[\frac{(1{,}250{,}000)(365)t}{(2.225 \times 10^9)}\right]$$

or:

$$A_x = 1.4466 + 0.3291t \qquad (9\text{–}25)$$

Interpolating from table 9–3 and assuming that the compressibility factor, z, is 1.00, the production forecast is set out in table 9–4 for intervals of six months. The results are also shown in figure 9–5 on semilogarithmic coordinates as a continuation of the first portion of the problem when the well had excess capacity and production was limited to 1,250 Mcfd. The

TABLE 9–4.
Production Forecast for a Well Producing Against a Constant Back Pressure of 1,000 psia

q_i = *1,250 Mcfd* — *Initial shutin pressure = 1,327.7 psia*
G_i = *2.225 × 10^9 cu ft* — *Constant back pressure = 1,000.0 psia*
n = *0.70* — z = *1.000*

t Time (yr)	A_x	x	Shutin Pressure (psia)	q at End of Time Period (Mcfd)	ΔGp Gas Produced in Interval (Mcf)	Gp Cumulative Gas Production (Mcf)	q Average Rate (Mcfd)
0.0	1.4466	1.3277	1,327.7	1,250	0	0	0
0.5	1.6112	1.2115	1,211.5	888	194,731	194,731	1,067
1.0	1.7757	1.1303	1,130.3	616	136,077	330,808	746
1.5	1.9403	1.0751	1,075.1	411	92,506	423,314	507
2.0	2.1048	1.0391	1,039.1	257	60,330	483,644	331

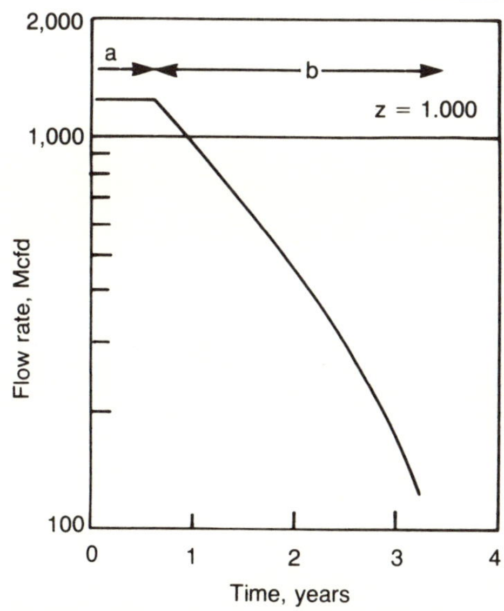

FIGURE 9–5 *Producing Rate for a Gas Well Producing a Constant Back Pressure (a) before Breaking Point and (b) after Breaking Point*

instantaneous rate of flow at the end of the time periods (fifth column of table 9–4) was calculated by means of equation 9–24:

$$q = q_i \left(\frac{x^2 - 1}{x_i^2 - 1} \right)^n \quad (9\text{–}24)$$

Equation 9–24 is inferred readily from the back-pressure relationship for gas wells.

Figure 9–5 shows that the producing rate for the well in this particular example decreases rapidly when the well produces against a constant pressure of 1,000 psia. In other words, production is limited by the high line pressure. In actual practice, a compressor probably would be installed as soon as the drop in producing rate became significant. Also, a comparison between figures 9–4 and 9–5 shows the effect of the premises as to back pressure conditions on the production forecast.

To the best of the author's knowledge, this is the first published detailed treatment of production forecasting for gas wells, even though forecasting has been done for years. As a matter of fact, the subject has received little attention in the literature except for the information in references 1, 2, and 3 given at the end of this chapter. The methods for forecasting given herein, except for the theoretical method in chapter 8, are of an empirical nature and evidence of their utility will be given in chapter 10. The methods were developed to give the practicing engineer a ready and straightforward method for forecasting gas production from wells. As a result of the lack of publications on forecasting methods, one can only speculate as to the actual methods in use in the industry. Probably, most large companies and consulting firms use stepwise material balance methods that require large, detailed computational models. However, the purpose here has been to present methods compatible with small computers.

SUMMARY

The methods presented in this chapter for forecasting production from gas wells are essentially empirical because they are based on the back-pressure equation. These methods do not account for gas produced when the well is producing under unsteady-state conditions. Also, the methods do not fully account for the change in the compressibility factor for the gas as the pressure decreases in the reservoir except as indicated for the method for estimating the time that a well will produce against a constant pressure when it has excess productive capacity. For the cases of a constant fraction of current shutin pressure and a constant back pressure, the changes in the compressibility factor can be treated adequately only by small steps with large computer facilities. Each of the methods requires an independent means for determining the gas in place available for producing the well. Unfortunately, the engineer may need to make the production forecasts at a time when very little is known about the extent of the reservoir and the volume of gas in place available for producing the well. Thus, ignoring the changes in the compressibility of the gas may be justified in many forecasts. The method for making a production forecast as presented in chapter 8 overcomes the difficulty caused by the changing compressibility of gas. If it is applied to a well in which the radius of drainage has not been established, it is then necessary to predict when the pressure disturbance will reach equilibrium with the pressure disturbance from other wells.

The methods have been presented without any proof of applicability to actual gas wells. Chapter 10 shall deal with production decline curves for gas wells and will present data on producing wells to show applicability.

PROBLEMS

1. For a gas well, use a stabilized back-pressure equation in which the difference in pressure squared is expressed in thousands (see figure 7–1 and table 7–3 for additional information) at the wellhead:

$$q_k = 42.96\ (\bar{p}_{ts}^{\ 2} - p_{tf}^{\ 2})^{0.647}$$

The current shutin wellhead pressure is 1,510 psia, and the gas in place available for production from the well at the current shutin is 2.1 billion cu ft.

a. Calculate the length of time that the well will deliver a steady production rate of 1,000 Mcfd against a back pressure of 915 psia.
b. Estimate the yearly production rates after (a) if the well continues to produce against the constant pressure of 915 psia.
c. Estimate the yearly production rates after (a) if the back pressure is reduced to the constant fraction of the current shutin pressure that existed at the time the well failed to make 1,000 Mcfd against the constant back pressure of 915 psia.

2. A well with a very thin section of relatively high-permeability reservoir produces 1,500 Mcfd against a back pressure of 850 psia at a current shutin wellhead pressure of 3,265 psia. Assume that n is 0.70. The pipeline will take any amount of gas up to about 10,000 Mcfd. The well is in a large common reservoir with several large-capacity wells, and the allowable for the subject well is subtracted from the total allowable to be divided among the capable wells. (In other words, the subject well has very little influence on the pressure production performance of the reservoir.) On the average, the wellhead shutin pressures of all the wells have been decreasing about 150 psi/year. It is proposed to install a compressor to reduce the pressure on the well to a constant 425 psia. Calculate the increase in gas production per year over the next 5 yr as a result of lowering the back pressure to 425 psia.

3. A gas well in the Texas Hugoton field has a current deliverability of 825 Mcfd with a shutin pressure of 88 psia. Its average allowable for the past year has averaged 72% of the deliverability, or 594 Mcfd, and is expected to remain constant in the future. Historically, the wellhead pressure has decreased about 10 psi/year. Calculate the time required for the well to produce at 100% of the deliverability. **Note:** Deliverability is defined in the Texas Hugoton field essentially as the rate of flow of a well against a pressure that is 80% of the wellhead shutin pressure. Assume that n is 0.85.

REFERENCES

1. M.J. Fetkovich. "Decline Curve Analysis Using Type Curves." *J. Pet. Tech.* (June 1980), p. 1065.
2. P.R. Stewart. "Low-Permeability Gas Well Performance at Constant Pressure." *J. Pet. Tech.* (September 1970), p. 1149.
3. J. Gurley. "A Productivity and Economic Projection Method—Ohio Clinton Sand Gas Wells." *J. Pet. Tech.* (November 1963), p. 1183.

10 Production Decline Curves for Gas Wells

In figures 8–1A and 8–1B and the discussion in chapter 8, the results of the theoretical development of a general production decline curve for gas wells producing against a constant back pressure were seen. In chapter 9, the problem was approached from the traditional empirical concept of gas well behavior. In the present chapter, the empirical approach shall continue to be covered and several actual well histories will be presented to show that the empirical approach is useful in practical natural gas engineering.

In chapter 9, it was convenient to present production forecasts for a hypothetical gas well on semilogarithmic coordinates with the production rate shown on the logarithmic coordinate and time shown on the normal coordinate. In figure 9–4 after the well reached the breaking point, the plot of the logarithm of rate against time was very nearly a straight line for a well producing against a pressure that was a constant fraction of the current shutin pressure. Actually, the calculated rate was slightly more optimistic than the straight line in that the calculated rates were larger than rates found by projecting the straight line. In figure 9–5 in which the well was producing against a constant back pressure, the calculated rates were less optimistic than a straight line.

This chapter will show how the ideas developed in previous chapters can be consolidated into generalized production decline curves for gas wells and will compare the theoretical decline curves with actual gas well behavior.

Equations 9–9, 9–10, 9–19, 9–20, and 9–22 can be rearranged so the dimensionless rate of flow, q/q_i, is a function of the dimensionless time, $q_i t/G_i$, for various values of the exponent n and the ratio, x_i, of the original shutin pressure to the constant back pressure imposed upon the well. As the original shutin pressure becomes large with respect to the constant back pressure, x_i^2 becomes large with respect to unity, and the problem becomes equivalent mathematically to the solution in which the well produces at a constant

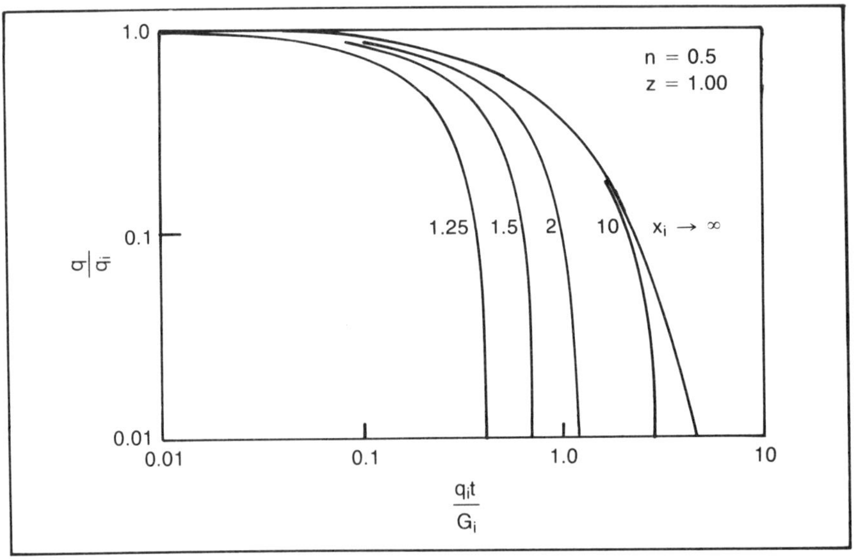

FIGURE 10–1 *Variation of Production Rate with Time for Wells Producing Against Various Constant Back Pressures,* n = *0.5*

fraction of the current open-flow potential or against a back pressure that is a constant fraction of the current shutin pressure. This is illustrated in figures 10–1, 10–2, 10–3, 10–4, 10–5, and 10–6 for values of the exponent n from 0.5 to 1.0. The dimensionless rate of flow, q/q_i, has been plotted against the dimensionless time, $q_i t/G_i$, on logarithmic coordinates on these six figures. Numerical values for the relationships are given in table 10–1 for the reader who wishes to make plots correspond to figures 10–1 to 10–6, inclusive.* The solutions given in the tables and figures were calculated from analytical solutions when available and otherwise from numerical solutions.

COMPARISON WITH THE THEORETICAL DECLINE CURVE

In chapter 8, a production forecast was made from the theoretical approach to the unsteady-state flow theory for gas wells, but in chapter 9 an empirical approach was developed. Using the same well example that was used in chapter 8, one can start at the stabilization point shown in table 8–9 and make an empirical production forecast for the same gas well using the

* The author recommends coordinates with at least 3-in. cycles.

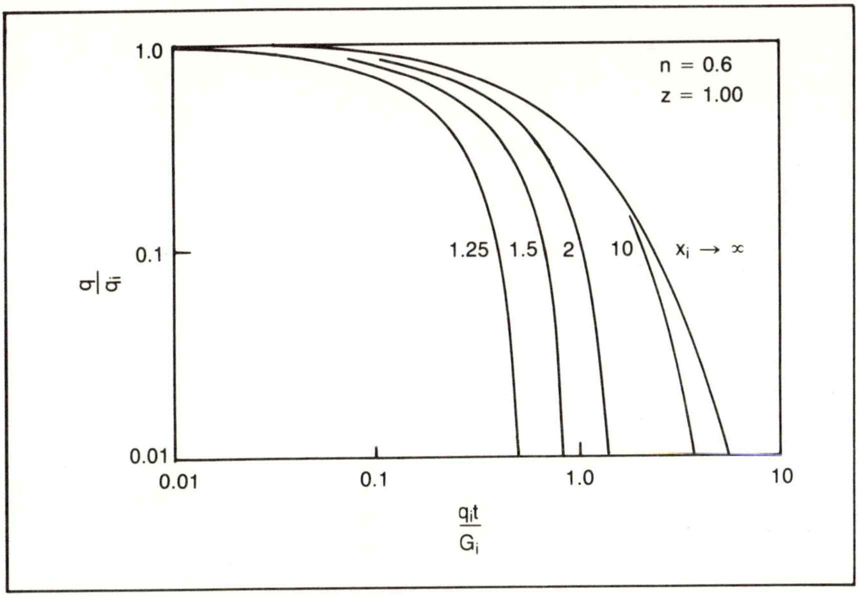

FIGURE 10–2 *Variation of Production Rate with Time for Wells Producing against Various Constant Back Pressures,* n = *0.6*

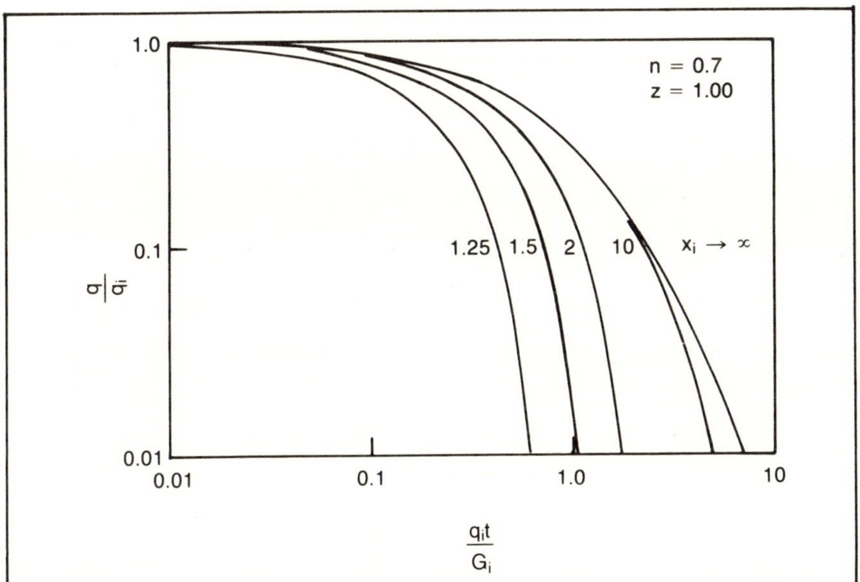

FIGURE 10–3 *Variation of Production Rate with Time for Wells Producing against Various Constant Back Pressures,* n = *0.7*

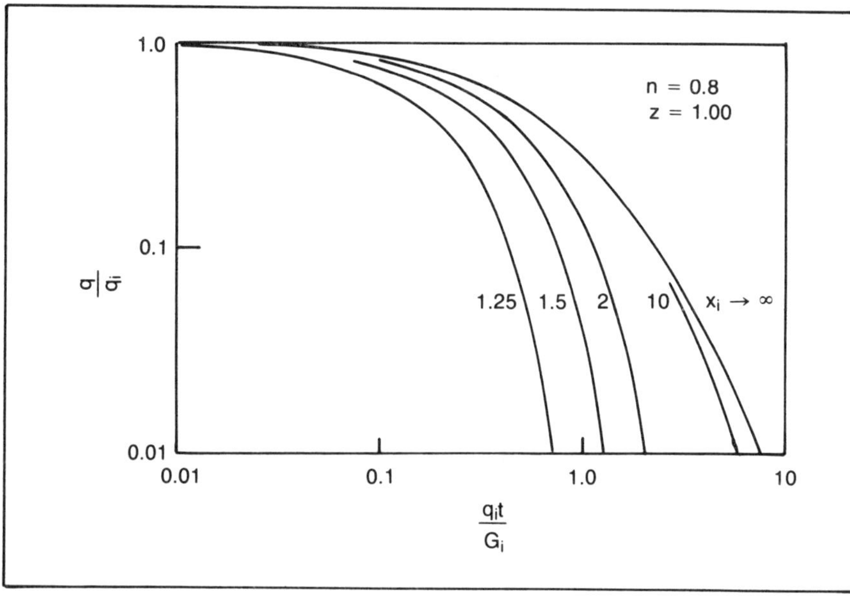

FIGURE 10–4 *Variation of Production Rate with Time for Wells Producing against Various Constant Back Pressures,* n = *0.8*

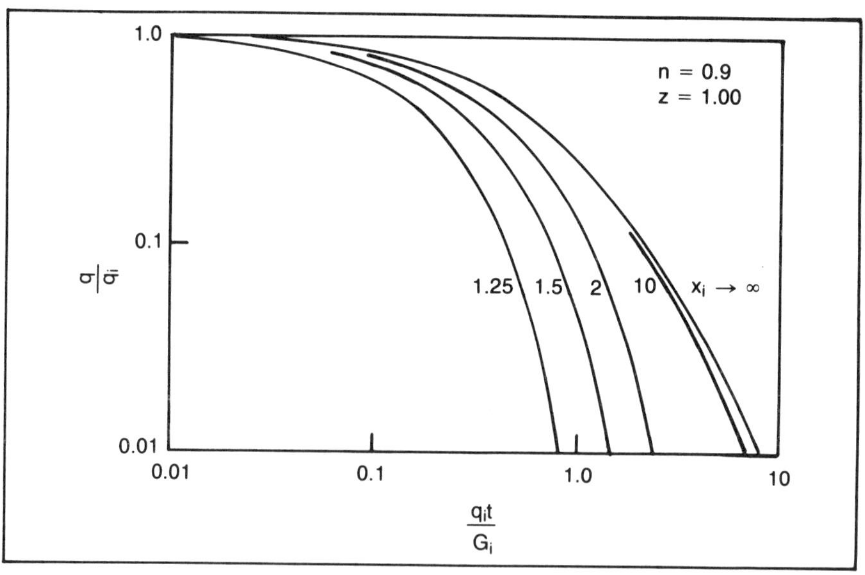

FIGURE 10–5 *Variation of Production Rate with Time for Wells Producing against Various Constant Back Pressures,* n = *0.9*

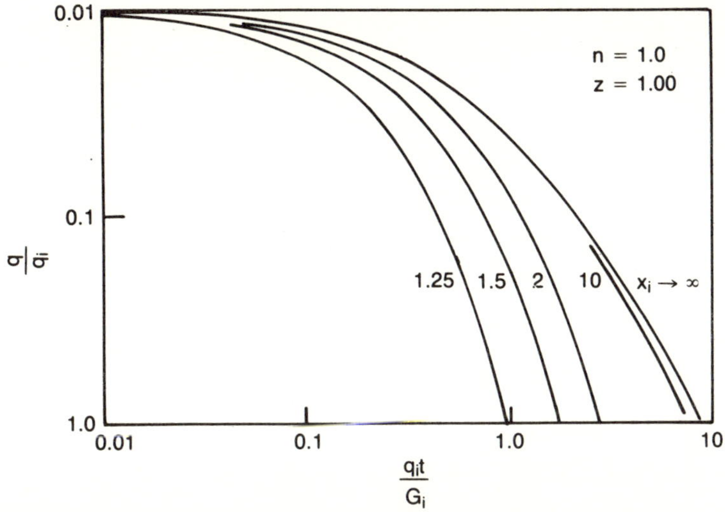

FIGURE 10–6 *Variation of Production Rate with Time for Wells Producing Against Various Constant Back Pressures,* n = *1.0*

same premises regarding the reservoir. Here, assume the well is producing against a constant back pressure of 662 psia in the wellbore at the level of the reservoir. Use the availability integral, A_x, from table 9–3 for an exponent, n, value of 0.70, but assume that the compressibility of the gas is 1.000 and use it as such in the material balance. The results are shown in figure 10–7 and are identified as the empirical method.

The difference in the forecast rates is not too bad for the 10 years after stabilization, and at 10 years the instantaneous predicted production rate is 344 for the empirical method versus 315 Mcfd for the theoretical method. Considering the assumptions involved in the empirical method, the chart reading of the theoretical method, and the widely different approaches to the problem, one could consider the agreement to be good. The theoretical method has a serious flaw in that it ignores the additional pressure loss caused by non-Darcy flow near the wellbore. However, the empirical method ignores the compressibility of the gas by assuming it to be 1.000. In the example selected, the exponent, n, is taken as 0.70, which indicates a large pressure drop caused by non-Darcy flow. This, in turn, results in a significant difference in the two methods.

TABLE 10–1.

Variation of Producing Rate with Time for Wells Producing against Various Constant Back Pressures and Various n Values, z = 1.00

Values of $(q_i t/G_i)$

	n = 0.5					n = 0.6				
q/q_i	$x_i \to \infty$	$x_i = 10$	2	1.5	1.25	$x_i \to \infty$	$x_i = 10$	2	1.5	1.25
1.00	0.000	0.000	0.000	0.000	0.000	0.000	0.000	0.000	0.000	0.000
0.95	0.051	0.051	0.038	0.028	0.018	0.043	0.042	0.032	0.024	0.015
0.90	0.105	0.104	0.078	0.057	0.037	0.089	0.088	0.066	0.048	0.031
0.80	0.223	0.221	0.162	0.118	0.075	0.189	0.187	0.139	0.101	0.064
0.70	0.357	0.352	0.254	0.181	0.114	0.306	0.303	0.220	0.158	0.100
0.60	0.511	0.504	0.353	0.249	0.154	0.444	0.439	0.312	0.221	0.138
0.50	0.693	0.682	0.462	0.320	0.196	0.612	0.603	0.417	0.292	0.180
0.40	0.916	0.899	0.580	0.394	0.238	0.825	0.811	0.539	0.370	0.225
0.35	1.050	1.027	0.643	0.433	0.260	0.956	0.938	0.608	0.413	0.250
0.30	1.204	1.174	0.709	0.472	0.282	1.111	1.088	0.683	0.460	0.276
0.25	1.386	1.344	0.776	0.512	0.304	1.300	1.267	0.766	0.510	0.304
0.20	1.609	1.546	0.846	0.552	0.326	1.538	1.492	0.858	0.564	0.334
0.15	1.897	1.794	0.918	0.593	0.349	1.859	1.785	0.962	0.624	0.367
0.10	2.303	2.105	0.991	0.634	0.371	2.339	2.197	1.082	0.692	0.405
0.07	2.659	2.332	1.036	0.659	0.384	2.788	2.546	1.165	0.739	0.430
0.05	2.996	2.502	1.066	0.676	0.393	3.238	2.853	1.228	0.775	0.449
0.03	3.507	2.685	1.096	0.692	0.402	3.970	3.254	1.301	0.815	0.471
0.015	4.200	2.830	1.118	0.705	0.409	5.068	3.662	1.368	0.853	0.492
0.010	4.605	2.879	1.126	0.709	0.411	5.772	3.835	1.395	0.868	0.500

Table 10–1, continued

q/q_i	n = 0.7					n = 0.8				
	$x_i \rightarrow \infty$	$x_i = 10$	2	1.5	1.25	$x_i \rightarrow \infty$	$x_i = 10$	2	1.5	1.25
1.00	0.000	0.000	0.000	0.000	0.000	0.000	0.000	0.000	0.000	0.000
0.95	0.037	0.037	0.028	0.020	0.013	0.032	0.032	0.024	0.018	0.012
0.90	0.076	0.076	0.057	0.042	0.027	0.067	0.066	0.050	0.037	0.024
0.80	0.165	0.163	0.121	0.088	0.056	0.145	0.144	0.107	0.078	0.050
0.70	0.268	0.265	0.194	0.140	0.089	0.239	0.236	0.173	0.125	0.080
0.60	0.393	0.388	0.279	0.199	0.125	0.352	0.348	0.252	0.108	0.113
0.50	0.548	0.540	0.379	0.267	0.165	0.495	0.488	0.346	0.245	0.153
0.40	0.748	0.737	0.500	0.346	0.212	0.683	0.673	0.464	0.324	0.199
0.35	0.874	0.860	0.571	0.392	0.238	0.804	0.791	0.536	0.370	0.227
0.30	1.026	1.007	0.652	0.443	0.267	0.951	0.935	0.619	0.423	0.257
0.25	1.215	1.189	0.744	0.499	0.300	1.136	1.115	0.716	0.485	0.292
0.20	1.460	1.423	0.852	0.565	0.336	1.381	1.350	0.835	0.558	0.334
0.15	1.799	1.741	0.983	0.642	0.379	1.728	1.681	0.986	0.649	0.384
0.10	2.327	2.219	1.147	0.737	0.431	2.286	2.200	1.190	0.768	0.451
0.07	2.845	2.662	1.274	0.809	0.471	2.851	2.708	1.358	0.865	0.504
0.05	3.384	3.089	1.379	0.868	0.503	3.459	3.229	1.507	0.950	0.550
0.03	4.308	3.729	1.513	0.943	0.543	4.541	4.080	1.715	1.067	0.614
0.015	5.800	4.520	1.655	1.022	0.586	6.384	5.295	1.960	1.204	0.688
0.010	6.819	4.920	1.721	1.059	0.606	7.706	5.999	2.086	1.274	0.726

Table 10–1, continued

q/q_i	$n = 0.9$					$n = 1.0$				
	$x_i \to \infty$	$x_i = 10$	2	1.5	1.25	$x_i \to \infty$	$x_i = 10$	2	1.5	1.25
1.00	0.000	0.000	0.000	0.000	0.000	0.000	0.000	0.000	0.000	0.000
0.95	0.029	0.029	0.022	0.016	0.010	0.026	0.026	0.019	0.014	0.009
0.90	0.060	0.059	0.045	0.033	0.021	0.054	0.054	0.040	0.030	0.019
0.80	0.130	0.129	0.096	0.070	0.045	0.118	0.117	0.087	0.064	0.041
0.70	0.215	0.212	0.157	0.114	0.072	0.195	0.193	0.143	0.104	0.066
0.60	0.319	0.315	0.229	0.165	0.104	0.291	0.288	0.210	0.152	0.096
0.50	0.451	0.445	0.318	0.226	0.141	0.414	0.409	0.294	0.210	0.132
0.40	0.628	0.620	0.432	0.303	0.187	0.581	0.573	0.404	0.285	0.177
0.35	0.743	0.732	0.503	0.350	0.215	0.690	0.681	0.472	0.331	0.204
0.30	0.885	0.870	0.586	0.404	0.246	0.826	0.813	0.555	0.385	0.236
0.25	1.065	1.046	0.686	0.468	0.284	1.000	0.983	0.656	0.450	0.274
0.20	1.306	1.280	0.812	0.547	0.328	1.236	1.213	0.785	0.532	0.321
0.15	1.655	1.615	0.977	0.648	0.385	1.582	1.548	0.960	0.641	0.383
0.10	2.228	2.158	1.212	0.788	0.464	2.162	2.102	1.220	0.798	0.471
0.07	2.826	2.710	1.419	0.909	0.530	2.780	2.682	1.459	0.940	0.550
0.05	3.483	3.298	1.612	1.019	0.591	3.472	3.319	1.692	1.075	0.624
0.03	4.689	4.319	1.897	1.181	0.679	4.774	4.470	2.054	1.283	0.738
0.015	6.832	5.930	2.266	1.389	0.792	7.165	6.419	2.558	1.568	0.893
0.010	8.428	6.965	2.471	1.503	0.854	9.000	7.768	2.857	1.735	0.984

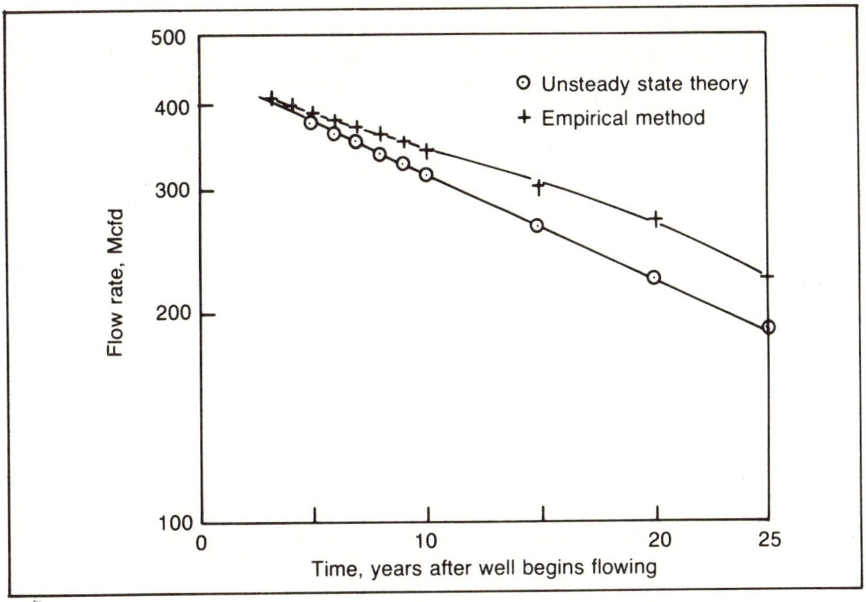

FIGURE 10–7 *Comparison of Production Forecasts by the Unsteady-State Theory with the Empirical Method*

PRODUCTION DECLINE BEHAVIOR OF ACTUAL GAS WELLS

The production for the complete life of a deep well in West Texas is shown in figure 10–8 in which production rate expressed in Mcf/month is shown plotted against time expressed in months on logarithmic coordinates. The well was remote from other producing wells, but fortunately it was close to a pipeline that operated at about 800 psig. The well produced from a reservoir with good permeability at a depth below 16,000 ft. It was opened into the line and checked periodically. After a producing life of 38 months, it was plugged and abandoned because operating it at a rate of less than 1,000 Mcf/month was uneconomical at the price then paid for gas in the area. Cumulative production to abandonment was 388,259 Mcf. The well produced small amounts of hydrocarbon liquid and water, and occasionally liquid accumulation in the wellbore interfered with the rate of production, which is apparent from examining the data in figure 10–8.

Figure 10–3 is used as a type curve and is shown in figure 10–8 as the dashed line coordinates and the two solid-line curves for $x_i = 2$ and 10. The pressure in the pipeline probably caused the well to perform as if it were

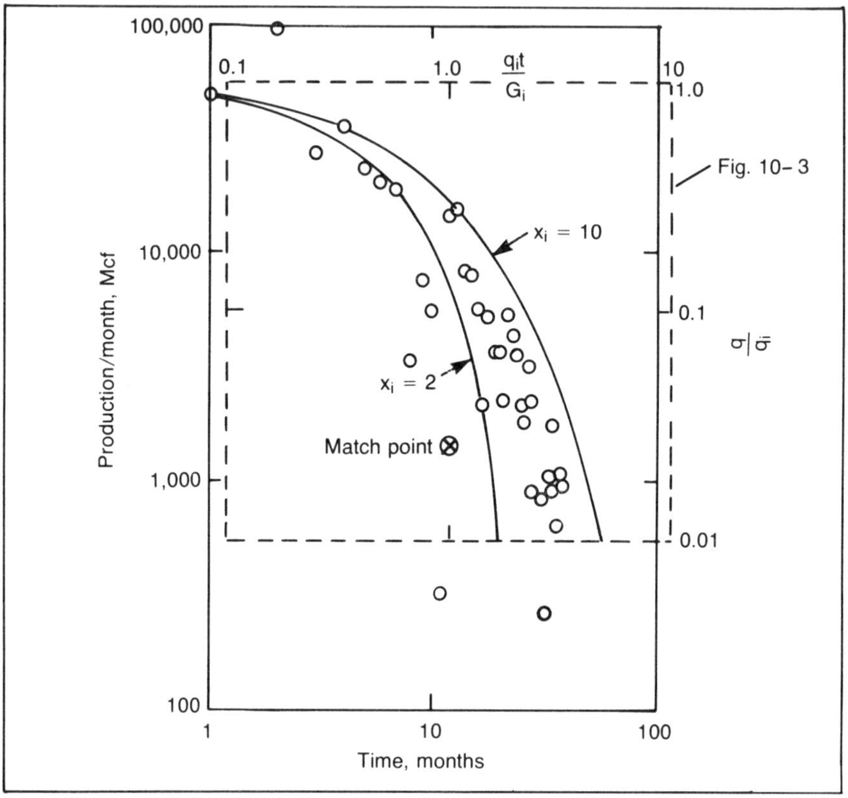

FIGURE 10–8 *Production Decline Behavior for a Deep, High-Pressure Well Completed in a Limited Reservoir, West Texas*

producing against a constant back pressure of $x_i = 7$ to 8. This indicates that the original shutin pressure was about 5,500 to 6,400 psi. The fit of the curves in figure 10–8 has been selected to give the most weight to the points on the right-hand side of the bulk of the data because these points were believed to be least influenced by liquid in the wellbore. The match point, as indicated in figure 10–8, is:

1,400 Mcf/month is equivalent to $q/q_i = 0.255$

12 months is equivalent to $q_i t/G_i = 1.00$

for which:

$q_i = 55{,}000$ Mcf/month

$G_i = 659{,}000$ Mcf

At abandonment, the well had produced about 59% of the initial gas in place. The well was completed in a limited reservoir and an offset well did not find the reservoir.

The production history of a well completed in a Mississippian reservoir in Oklahoma is shown in figure 10–9. The original shutin pressure at the bottom of the hole was about 2,070 psia. The pipeline pressure at the wellhead varied from 440 to 470 psig, averaging about 450 psig over the 75-month producing life of the well. However, figure 10–9 shows only the first 38 months of production because production thereafter dropped to 500 Mcf or less per month. These latter production data points have not been shown because the performance of the well was believed to have been damaged by the presence of liquids in the wellbore. Total production to abandonment was 155,459 Mcf and 1,934 bbl of condensate. Water production was estimated to be about equal to that of the condensate.

In fitting the type curve on figure 10–9, most of the weight for fixing the position of the type curve was given to the points indicating the best performance for the well—that is, the largest producing rates for the longest time on the time coordinate. For the match point indicated in figure 10–9:

830 Mcf/month is equivalent to $q/q_i = 0.10$

38 months is equivalent to $q_i t/G_i = 0.98$

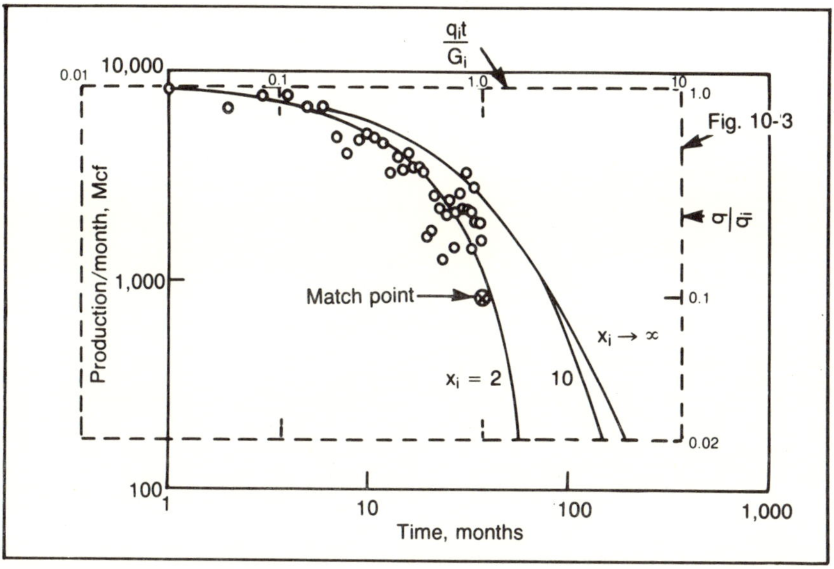

FIGURE 10–9 *Production Decline Behavior for a Mississippian Well in a Limited Reservoir, Oklahoma*

for which:

q_i = 8,300 Mcf/month
G_i = 322,000 Mcf

A number of shutin pressures (seven-day duration) were taken during the life of the well, and it has been possible to plot the relationship between p_R/z and cumulative production as shown in figure 10–10. If the most favorable point (figure 10–10) is taken as an indication of the proper relationship between p_R/z and cumulative production, the resultant original gas in place is 262,000 Mcf. The relationship between p_R/z and cumulative production for the well is by no means a straight line. If one uses the original gas in place as indicated by the production decline curve in figure 10–9, one can draw a straight line on figure 10–10 from the initial p_R/z of 2,501 to 322,000 Mcf. The result bears a believable relationship to the p_R/z data when one realizes that the performance of the well was probably influenced by the presence of liquids in the wellbore and unsteady-state effects. One cannot consider the seven-day pressure measurements shown in figure 10–10 as completely built-up pressure; therefore, the true p_R/z versus cumulative production curve should lie above and to the right of the data points in figure 10–10.

For the third example, the production history of a gas well in central Pennsylvania is shown in figure 10–11. The well discovered a long, narrow,

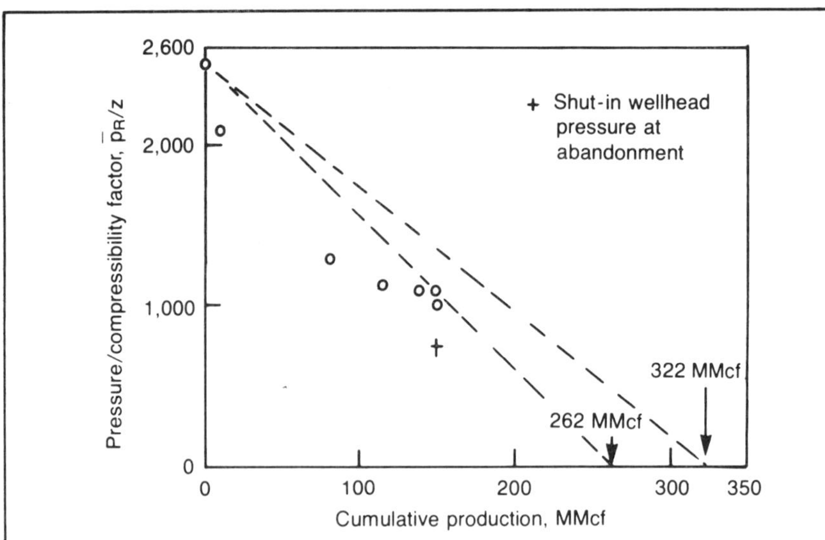

FIGURE 10–10 p/z *against Cumulative Production for a Mississippian Well in a Limited Reservoir, Oklahoma*

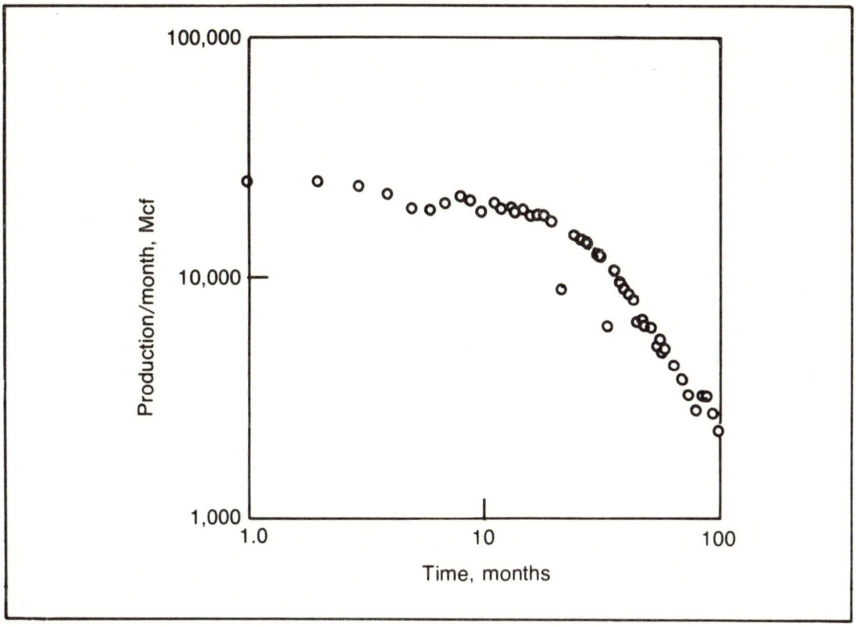

FIGURE 10–11 *Production History for a Gas Well in Pennsylvania*

and relatively dry gas reservoir in which several additional wells were eventually drilled. Compressors were installed and the back pressure on the well was lowered in an effort to maintain producing rates. The data shown in figure 10–11 covers the virtual life of the reservoir and the wells over a period of about 100 months. However, not all of the monthly rates are given in the figure in order to reduce the congestion of data points. Monthly production rates are shown up to 20 months, production rates for every other month are given from 20 to 60 months, and rates are for every fifth month from 60 months on for the life of the well.

A comparison of the data in figure 10–11 with the empirical production decline curves in figures 10–1 to 10–6 indicates that the data in figure 10–11 cannot be fitted in entirety to a single curve. However, the producing life up to about 25 months can be fitted very well to the limiting curve as x_i approaches infinity in figure 10–3 as shown in figure 10–12. In other words, the well acted as if it were producing against a changing back pressure kept at a constant fraction of its current shutin pressure for its early producing life. The match point, as indicated in figure 10–11, is:

16,300 Mcf/month is equivalent to $q/q_i = 0.58$

38 months is equivalent to $q_i t/G_i = 0.69$

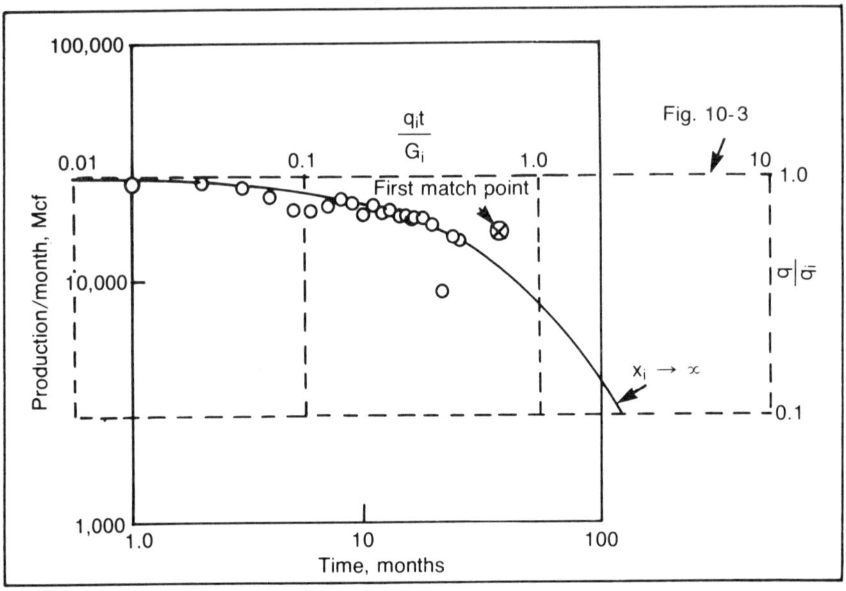

FIGURE 10–12 *Production Decline Behavior in Early Producing Life for a Gas Well in Pennsylvania*

for which:

q_i = 28,100 Mcf/month
G_i = 1,550,000 Mcf

If one accepts that the production decline relationship could change with time during the life of the well, one makes a second comparison using figure 10–3, which is shown in figure 10–13. Here, the fit to the $x_i \rightarrow \infty$ curve is considered good, but the most weight has been given to the production points that lie above and to the right of the bulk of the points. This is based on the philosophy that little happens to a gas well to make it perform better than it should. The match point for the performance during the later producing life is:

6,600 Mcf/month is equivalent to q/q_i = 0.172
84 months is equivalent to $q_i t/G_i$ = 2.45

for which:

q_i = 38,400 Mcf/month
G_i = 1,290,000 Mcf

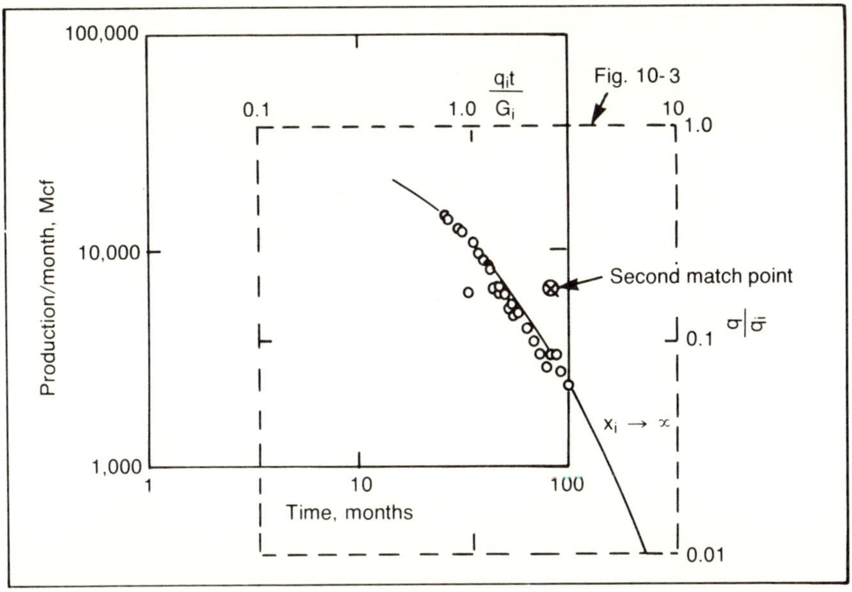

FIGURE 10–13 *Production Decline Behavior in a Late-Producing Life for a Gas Well in Pennsylvania*

The second value of the initial gas in place is 260,000 Mcf less than the indicated early life value of 1,550,000 Mcf, or a 17% decrease.

Recalling that the well in this example was the discovery well in the reservoir and that other wells were completed in the same reservoir and placed on production, it is logical to assume that the completion of other wells in the same reservoir would reduce the amount of gas in place that was available for production from the discovery well in the reservoir. The timing of the occurrence of the reduction of the amount of gas in place would depend on reservoir characteristics such as permeability, porosity, and shape. The size of the reduction would depend, for the most part, on the producing rates of the wells involved and the pressures in the reservoir.

It is well worth noting that the performance of the well in the third example was that of a well producing at a constant fraction of its current open flow potential. This is the same problem mathematically as for a well producing against a back pressure that is a constant fraction of the current shutin pressure. This performance contrasts with the first and second well examples in which the wells produced against a nearly constant back pressure.

For the fourth example of a production decline history, consider a gas well in the Texas Hugoton field. The well was placed on production in the late

1940s and has produced gas since that time. Production for the past 8 years follows:

Year	*Production (Mcf)*
1972	271,588
1973	327,535
1974	362,905
1975	331,972
1976	317,555
1977	250,887
1978	233,503
1979	211,953

Although the production history is a limited amount of information, use type production decline curves to determine the amount of gas in place. Later, additional information on the performance of the well will be introduced to check the conclusions derived from the production decline study.

The first conclusion to be reached from the production history is that the well is a large-capacity producer, considering the present low pressure prevalent in the Texas Hugoton field. Also, note that the producing rate increased to a maximum in 1974 and declined thereafter. Other things being equal, one must conclude that the well reached some sort of breaking point. Since the Texas Hugoton field is prorated, the market demand for gas after being reduced by the amount of gas to be produced by all wells not capable of producing a full allowable is divided among the capable wells. This means that as pressures in the field decrease, the productivity of the wells decrease and the allowable assigned to the capable wells increases. This is believed to be the reason for the increase in production to 1975 for the well under consideration.

If one uses production for 1974 through 1979 and makes a log-log plot as shown in figure 10–14 by the data points, an excellent fit can be made to the $x_i \rightarrow \infty$ curve in figure 10–4 as depicted in figure 10–14. Note that the yearly production is plotted at the midyear point. The match point is:

145,000 Mcf/year is equivalent to $q/q_i = 0.35$

5.7 years is equivalent to $q_i t/G_i = 0.50$

for which:

$q_i = 414{,}000$ Mcf/year

$G_i = 4{,}720{,}000$ Mcf

Since the analysis started with the production for 1974, the production for the period 1974–1979 inclusive must be subtracted from 4,720,000 Mcf to find

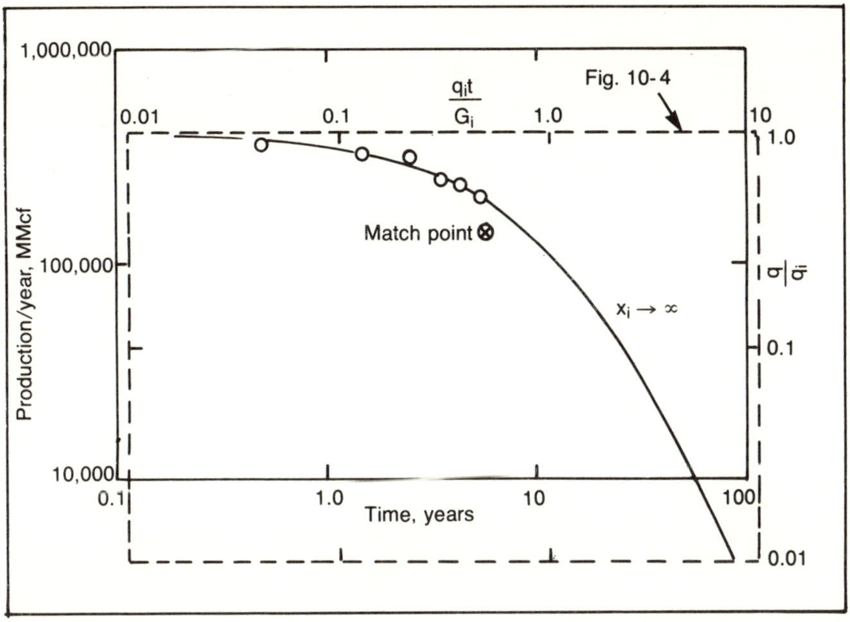

FIGURE 10–14 *Production Decline Behavior for a Well in the Texas Hugoton Field*

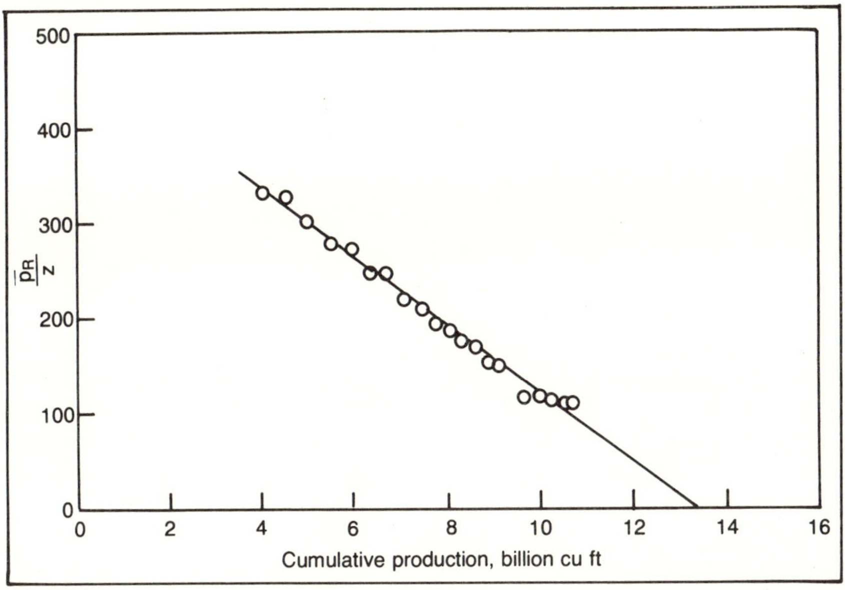

FIGURE 10–15 p_R/z *and Cumulative Production for the Gas Well in the Texas Hugoton Field*

the gas in place available for production from the well at the end of 1979. The result is 3,010,000 Mcf.

In order to check this result, information on test pressures and yearly production is obtained, p_R/z is calculated, and accumulative production against p_R/z is plotted in figure 10–15. The solid line in the figure represents the least squares fitted to the data points shown. The line indicates the gas in place at the end of 1979 to be 2,550,000 Mcf but, as so often happens, there is a dilemma. The last five data points in figure 10–15 shown an awkward trend. As noted previously from the production data, the example well underwent a change in production trend for the past six years. Also, pressure measurements (taken at the wellhead) may have been influenced by liquid standing in the wellbore. Therefore, the gas in place determination as of the end of 1979 of 3,010,000 Mcf by the production decline method is probably as likely as that of 2,550,000 Mcf from the more conventional $\bar{p}_R/z$ cumulative production method.

SUMMARY

This chapter has attempted to establish the utility of using the empirical production decline curves develoed in chapter 9 to make production forecasts and to estimate the gas in place available for producing a well. Type curves are presented in figures 10–1 through 10–6 for making comparisons with the production performance of actual wells. Also, the values used to make the plots for the type curves are presented in table 10–1 so the reader can construct his personal type curves. However, log-log coordinates with cycles no less than 3 in. on a side are recommended for the plots.

The comparison of the empirical decline method with the theoretical method of chapter 8 for production forecasting is believed to be good, considering the widely different approaches to the problem and the flaws in each method. However, the empirical method seems to provide the most optimistic answer for the one production forecast presented herein.

Four comparisons of the empirical production decline curves are made using the production histories of actual wells. Three of the four production histories are for the entire productive life of the wells, which provides a comparison over the complete length of the type curve. The fourth comparison is made over a small time interval in the productive life of a well with a large amount of gas in place that is available for production from the well. Where the data are available, the gas in place is estimated by plots of p_R/z against cumulative production. Here the comparisons have been excellent.

The examples, except for the Oklahoma well, are from wells that stabilized quickly and were initially capable of producing gas at relatively large rates of flow. Do not attempt to apply the empirical type curves to wells with producing rates affected by unsteady-state conditions.

Unfortunately, the technical literature on production decline curves and production forecasting is practically nonexistent except for the three references given at the end of chapter 9. It is hoped that the material presented in chapters 9 and 10 will stimulate interest in the subject and encourage future publications.

A word of caution is necessary regarding the unit of time used in reporting gas production as practiced in some areas: Gas measurements in many places are recorded on eight-day charts, so seven-day recordings can be made on an eight-day chart with some flexibility in installing and removing the charts. Since most gas production reporting is done on a monthly basis and seven-day recordings are not compatible with monthly reporting, the least-costly method for making them compatible is arbitrarily to use 28 days in the first and second "months" and 35 days in the third "month" Thus the calendar matches the arbitrary schedule every 91 days. Under this system, a gas well will show an exceptionally good quantity of production every third month. This can be a puzzling circumstance to anyone not aware of the system for making weeks fit nicely into months.

PROBLEMS

1. The production history for a large-capacity gas well follows. Estimate the initial gas in place and calculate the fraction of gas in place produced at the end of the production history.

Month	*Production (Mcf)*	*Month*	*Production (Mcf)*
1	137,792	25	38,195
2	156,079	26	37,856
3	132,681	27	30,186
4	136,731	28	31,671
5	120,615	29	28,547
6	115,589	30	24,549
7	103,547	31	24,018
8	101,154	32	20,904
9	90,616	33	19,660
10	78,505	34	15,847
11	74,353	35	21,624
12	68,654	36	20,655
13	68,500	37	18,012
14	62,803	38	17,638
15	57,499	39	16,076
16	56,763	40	18,732

(continued)

Month	*Production (Mcf)*	*Month*	*Production (Mcf)*
17	57,599	41	18,244
18	56,193	41	15,752
19	50,461	43	14,640
20	49,463	44	16,198
21	48,055	45	18,392
22	43,747	46	14,758
23	43,482	47	13,082
24	38,089	48	13,692

2. Make similar calculations as in problem 1 for the following production history for an actual gas well.

Month	*Production (Mcf)*	*Month*	*Production (Mcf)*
1	1,302	19	6,209
2	8,400	20	5,855
3	10,838	21	4,933
4	11,559	22	4,569
5	10,659	23	4,227
6	12,159	24	3,639
7	11,949	25	3,510
8	12,563	26	3,135
9	12,283	27	3,125
10	12,746	28	2,664
11	12,398	29	2,631
12	12,125	30	2,440
13	12,439	31	2,385
14	12,146	32	2,148
15	11,711	33	3,017
16	10,254	34	1,719
17	8,152	35	1,673
18	7,866	36	1,562

3. Examine the four examples given in chapter 10 and your results for problems 1 and 2 to determine the least amount of production history that would have been necessary to derive about the same amount of gas in place as was determined using the given production history.

11 Sizing Flow Strings For Gas Wells

A flow string, or the tubing installed inside the casing of a gas well, can serve at least four important functions. First, if the tubing is installed with a packer near the bottom of the hole, the tubing protects the casing from the pressures of the well fluids in the tubing. Second, it protects the casing from the corrosive effects of the flowing fluids. Third, if sized and set properly, the use of tubing can keep the wellbore free of hydrocarbon liquid and water that otherwise might stand in the wellbore. Fourth, the conductor string must be large enough to permit the well to meet the largest demand for gas within the restraint imposed by the back pressure at the wellhead. The fourth listed function is, within various limits, the most important function. The first two functions are principally subject to mechanical and metallurgical design and will not be discusssed here.

TUBING SIZES REQUIRED FOR VARIOUS RATES OF FLOW

The production of gas from a reservoir into a pipeline or gathering system involves the pressure loss from the reservoir into the wellbore, the pressure loss from lifting the gas from the bottom of the well to the surface, overcoming friction in the flow string, and the pressure loss caused by cooling the gas from the reservoir to surface temperatures. The two pressure losses caused by lifting the gas and overcoming friction will be combined to show the effect of the size of the flow string upon the capacity of a well to deliver gas at the wellhead. Assuming that the mechanical features of the tubing design present no difficulties, the sizing is reduced to a balance between well productivity, market demand for gas from the well, lifting liquids from the wellbore, and costs. The first step in sizing tubing is to establish the back-pressure curve for bottom-hole conditions. The bottom-hole shutin pressure, $\bar{p}_R$, and selected rates of flow with corresponding flowing pressures, p_{wf}, should be converted to wellhead pressures using the methods outlined in chapter 5 for several tubing sizes. The next step is to make production forecasts as outlined in chapters 8 and 9. The production forecasts should be evaluated economically to determine which tubing size gives the best return on investment. However, experience will show that, for wells of moderate productive capacity in known areas, it is not necessary to make the production forecasts and economic analyses, but where the size of tubing influences the sizes of the drilled hole and casing, the complete analysis should be made.

The results of this procedure for tubing design are illustrated for a hypothetical gas well. The shutin pressures and back-pressure characteristics for the well are given in table 11–1 and in figure 11–1. The shutin pressure and the flowing pressures at the bottom of the hole (depth = 10,250 ft) are converted to wellhead pressures by the methods given in chapter 5 for the five sizes of tubing. Wellhead back-pressure data are calculated as shown in table 11–2 and back-pressure performance curves at the wellhead are given in figure 11–2. If a well with a shutin pressure of 3,000 psia at the bottom of the hole and a corresponding wellhead pressure of 2,437 psia is opened to flow against a wellhead pressure of 1,000 psia, the rates of flow would vary from 11,500 Mcfd for 2⅜-in. OD tubing to 56,700 Mcfd for 4½-in. tubing (table 11–3). Also, flow rates for other pressure conditions are given in table 11–3.

Note that a large-capacity well has been selected for the hypothetical well to demonstrate flow capacities for a wide range of tubing sizes. If the

TABLE 11–1.

Shutin Pressures and Back Pressure Characteristics for a Hypothetical Gas Well

$\bar{p}_{ts}$ (psia)	$\bar{p}_R$ or p_{wf} (psia)	$\bar{p}_R^2 - p_{wf}^2$ (M)	q_k (Mcfd)
2,437	3,000	—	0
—	2,990	59.9	3,000
—	2,972	167.2	6,000
—	2,943	338.8	10,000
—	2,844	911.7	20,000
—	2,719	1,607.0	30,000

γ_g = 0.625, H = 10,250 ft, bottom-hole temperature = 251°F, and wellhead temperature = 130°F

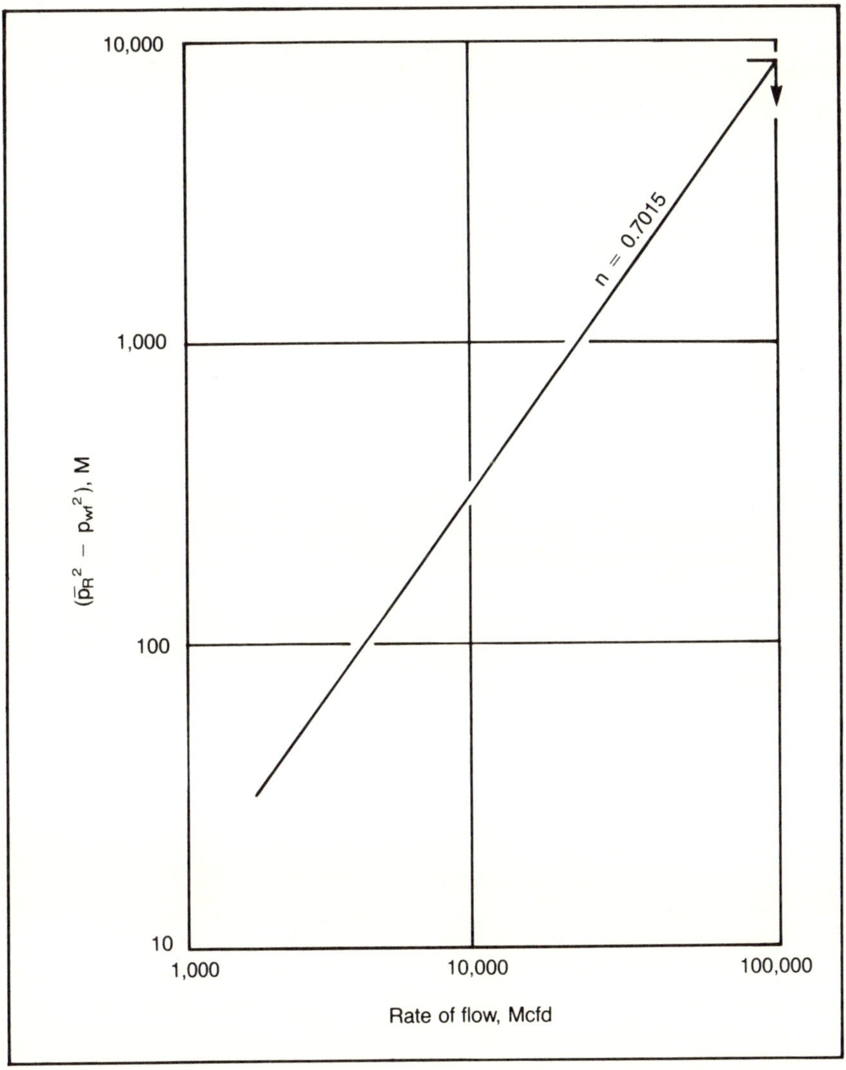

FIGURE 11–1 *Bottom-Hole Back-Pressure Performance for a Hypothetical Well*

hypothetical well had a contract obligation or an allowable of 3,000 Mcfd, an examination of table 11–3 shows that 2⅜-in. OD tubing would be adequate to produce 3,000 Mcfd against a wellhead pressure of 1,000 psia until the shutin wellhead pressure was 1,166 psia. In this case, if the demand for gas were

TABLE 11–2.

Differences of Pressure Squared (Thousands) for Wellhead Conditions for Various Sizes of Tubing

$(\bar{p}_{ts}^2 - p_{tf}^2)$ *(thousands)*

q_k (Mcfd)	OD (in.) = 2⅜ ID = 1.995	2⅞ 2.441	3½ 2.992	4 3.476	4½ 4.082
3,000	360	152	78	58	—
6,000	1,386	557	265	182	142
10,000	3,781	1,459	652	420	309
20,000	—	—	2,308	1,382	940
30,000	—	—	5,064	2,815	1,818

3,000 Mcfd, 2⅜-in. tubing would be adequate. A demand for 30,000 Mcfd gas could be met by 3½-in. tubing for only a short time or until the wellhead shutin pressure was reduced to 2,428 psia from 2,437 psia. However, the demand for 30,000 Mcfd could be met with 4½-in. tubing against a 1,000 psia pressure at the wellhead until the shutin wellhead pressure was reduced to 1,466 psia. The minimum wellhead shutin pressures given above were calculated by equation 11–1 which follows:

$$\bar{p}_{ts} = [p_{tf}^2 + (q/C)^{1/n}]^{0.5} \tag{11–1}$$

where the symbols are for wellhead conditions.

Equation 11–1 was used to calculate the minimum shutin wellhead pressures required to produce the designated flow rate against a back pressure of 1,000 psia for the hypothetical well. The results are illustrated in figure 11–3. For example, if the demand for gas were 10,000 Mcfd, the well could produce that quantity of gas through 2⅜-in. OD tubing until the shutin wellhead pressure was 2,185 psia, through 2⅞-in. until the pressure was 1,568 psia, and through 3½-in. until the pressure was 1,293 psia. The hypothetical well could not produce 15,000 Mcfd through 2⅜-in. OD tubing if the initial shutin wellhead pressure were 2,437 psia or 3,000 psia at the bottom of the hole. The incentive for installing tubing with larger than 3½-in. OD tubing is small if the demand for gas does not exceed 10,000 Mcfd. However, demands for gas of 15,000 Mcfd or more require economic analysis to make the decision to install tubing larger than 3½-in. OD.

As experience is gained in developing a reservoir and the market possiblities for the gas have been explored, it is probably necessary to make a

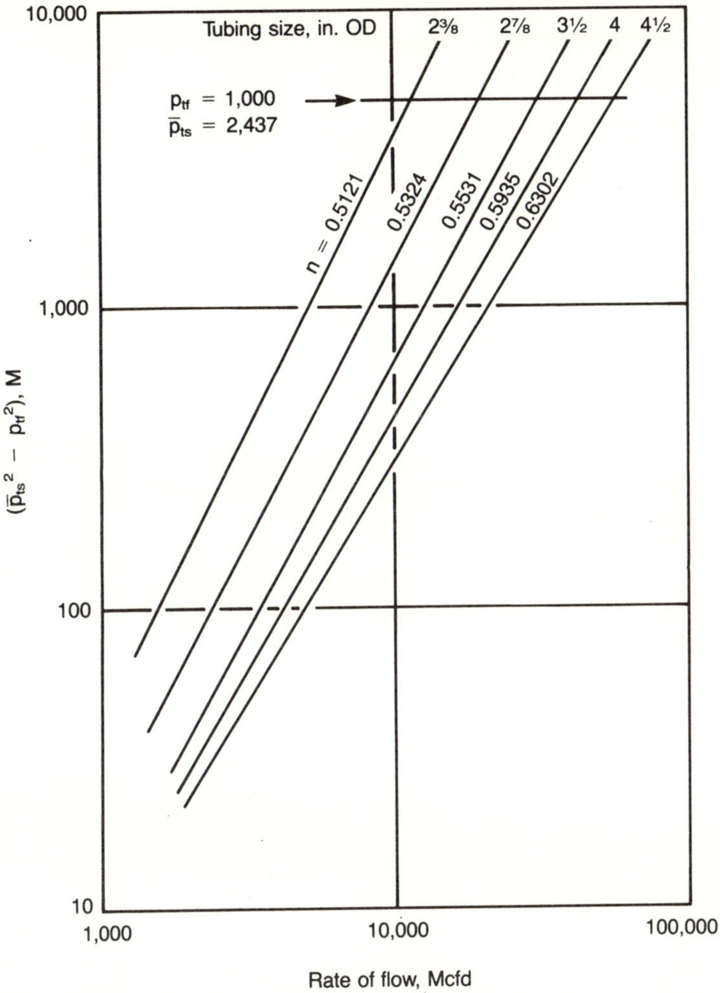

FIGURE 11–2 *Effect of Tubing Size on Availability of Gas at Wellhead for a Hypothetical Well*

complete analysis to determine the proper tubing size for only one well in the reservoir. Studing figure 11–3 shows that the choice of size for average gas demands of 5,000 to 10,000 Mcfd is limited to 2⅜, 2⅞, or 3½-in. tubing. Only an exceptional circumstance, such as high peak demands for gas, justifies the 3½-in. tubing.

TABLE 11–3.
Flow Rates for Hypothetical Well Through Various Sizes of Tubing

Tubing Size (OD, in.)	$\bar{p}_{ts}$ = 2,437 p_{tf} = 1,000 Flow Rate (Mcfd)	$\bar{p}_{ts}$ = 1,500 p_{tf} = 1,000 Flow Rate (Mcfd)	$\bar{p}_{ts}$ = 1,200 p_{tf} = 1,000 Flow Rate (Mcfd)
2⅜	11,500	5,700	3,300
2⅞	19,200	9,200	5,300
3½	30,100	14,100	7,900
4	42,400	18,800	10,100
4½	56,700	23,900	12,400

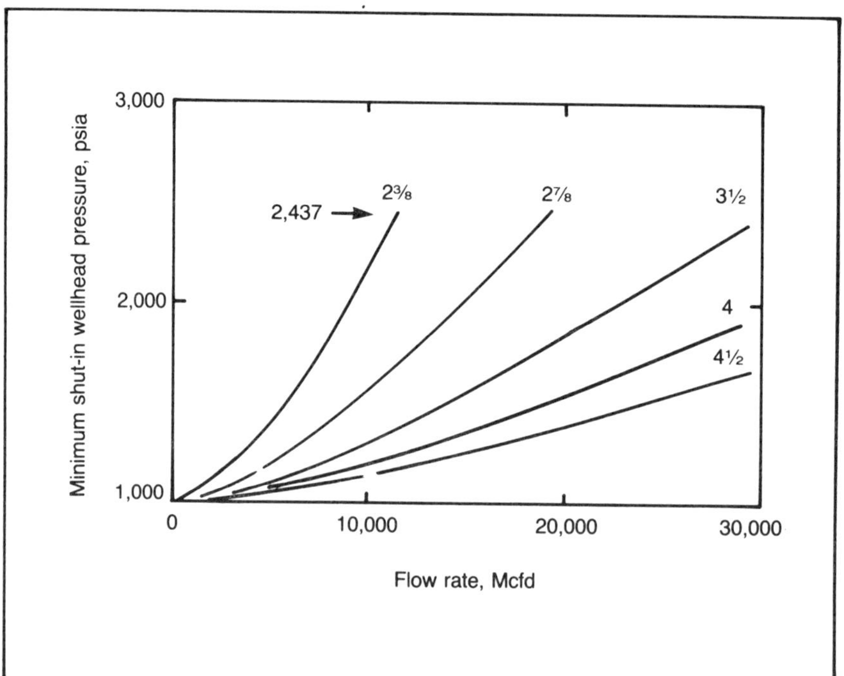

FIGURE 11–3 *Minimum Shutin Wellhead Pressure Required to Produce Indicated Flow Rates against 1,000-psia Back Pressure Through Various Sizes of Tubing*

SIZING TUBING TO LIFT LIQUIDS AT VARIOUS FLOW RATES

If meeting the market demand for gas is the only requirement, tubing for gas wells can be sized by installing the largest tubing the casing and workover procedures will accommodate. But the tubing should also be sized to lift liquids from the well. Practically all gas wells produce some liquids by either or both of two mechanisms. Liquids, either hydrocarbon or water, may enter the wellbore along with the gas, or they may condense out of the gas as it moves up the tubing string. If the velocity is too low to carry the liquids up the tubing, they will move down the tubing against the flow of the gas and accumulate in the wellbore.

The necessity for keeping liquids moving out of the wellbore and flowstring is often neglected through oversight and failure to realize the problems that arise when a gas well is allowed to accumulate liquids in the wellbore. The obvious reasons for liquid removal are that the column of liquid imposes an additional back pressure on the formation with adverse effects on productivity and the possibility that enough liquid to kill the well will accumulate. This happens often in low-pressure wells that produce large quantities of formation water. A more subtle reason for keeping liquids from the wellbore is that liquids standing in a wellbore opposite the producing formation tend to increase the liquid saturation in the vicinity of the wellbore and thereby decrease its permeability to gas. The net result is damage to the productive capacity of the well.

Experience with the low-pressure wells in the West Panhandle and Hugoton fields has shown that a velocity of 5 to 10 ft/sec is necessary to consistently remove hydrocarbon liquids, and a velocity of 10 to 20 ft/sec is required for water. The problem was described by Duggan, who estimated that a velocity of 5 ft/sec in tubing is required to remove hydrocarbon liquids.[1] The problem was analyzed theoretically by Turner, Hubbard and Dukler.[2] They presented a theoretical equation for calculating the minimum velocity for removing hydrocarbon condensate and water from the flow strings of gas wells. The equation indicates that the minimum velocity is dependent upon the interfacial tension between the liquid and the gas raised to the 0.25 power and the difference between liquid density and gas density to the 0.25 power and is inversely proportional to the square root of the gas density. After an adjustment to improve agreement with experimental date on actual gas wells and substitution of average values for the densities of water and condensate, Turner et al. presented the following equations:

$$u_g(water) = \frac{5.62(67 - 0.0031p)^{1/4}}{(0.0031p)^{1/2}} \qquad (11\text{–}3)$$

$$u_g(condensate) = \frac{4.02(45 - 0.0031p)^{1/4}}{(0.0031p)^{1/2}} \quad (11\text{–}4)$$

where $0.0031p$ reprsents the density of the gas. This term in equations 11–3 and 11–4 can be replaced with equation 2–14 of chapter 2, giving the density of the gas phase:

$$\rho_g = 2.6988 \frac{p\gamma_g}{zT} \quad (2\text{–}14)$$

However, most applications to gas well problems do not require the more-accurate gas densities calculated by equation 2–14.

It is of interest to determine the impact of equations 11–3 and 11–4 on gas well operation as illustrated in figure 11–4 in which the minimum velocity required to lift condensate and water are shown as functions of pressure. Figure 11–4 shows that the lower the pressure, the larger the velocity required to lift liquids. This is unfortunate because low-pressure wells are the most likely wells to have liquids in the wellbore and are the least capable of producing the higher velocities. Eventually, it is necessary to install mechanical pumps in low-pressure wells to keep them free of liquids.

The velocity calculated from equations 11–3 and 11–4 can be converted to flow rates using equation 5–12, which has been rearranged as follows:

$$q_m = 0.01673 \frac{pd^2u}{zT} \quad (11\text{–}5)$$

or:

$$q_m = 3.067 \frac{pAu}{zT} \quad (11\text{–}6)$$

where:

A = cross-sectional area sq ft

Returning to figure 11–3 the region on the flow capacity curves where the velocity of the gas at 1,000 psia is not enough to lift liquids from the well can be calculated. At 1,000 psia, the minimum velocities required are 5.81 ft/sec

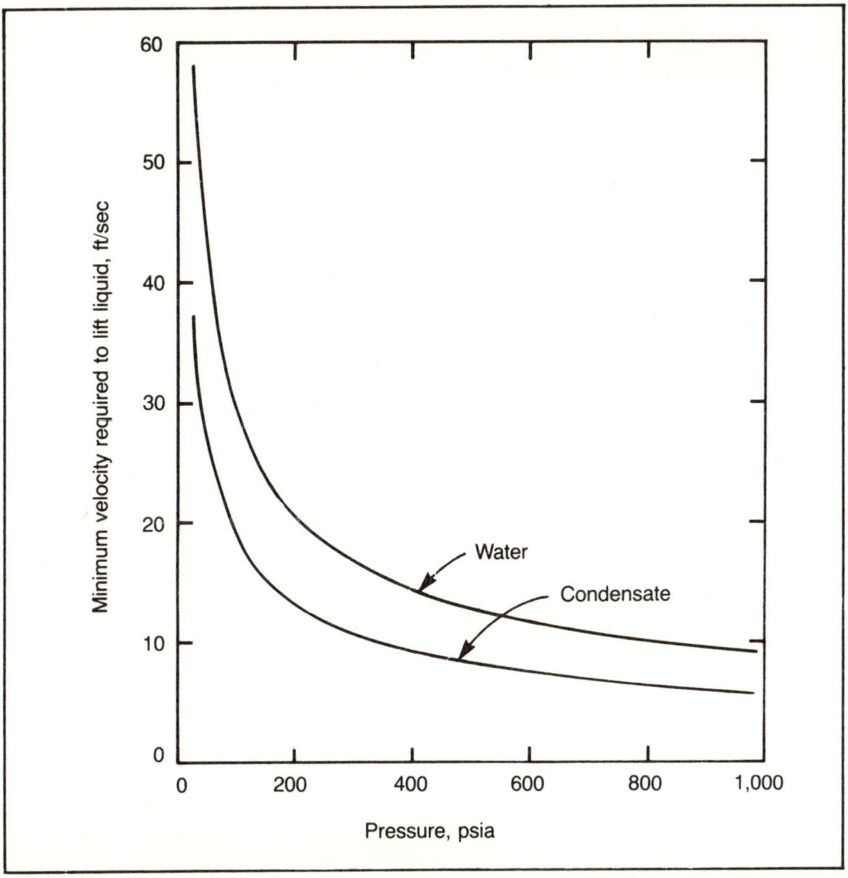

FIGURE 11–4 *Minimum Gas Velocity Required to Lift Hydrocarbon Liquid and Water in a Gas Well*

for hydrocarbon condensate and 9.02 ft/sec for water. Substitution into equation 11–5 give the minimum rates of flow required for lifting water and condensate in the various sizes of tubing (see table 11–4). **Note:** The minimum rates of flow given in table 11–4 are independent of the capacity of a well to produce gas. Thus, a well completed with 4½-in. tubing as casing but without a smaller flow string installed would require a rate of flow of 4.77 MMcfd at a pressure of 1,000 psia, a temperature of 130°F, and a specific gravity of 0.625 to keep the wellbore free of water. At lesser rates of flow, water would accumulate in the wellbore.

Equally important to the proper sizing of the tubing is the placement of the bottom of the tubing string. In order to effectively lift water from the

TABLE 11–4.
Minimum Rates of Flow Required to Lift Water and Condensate in Various Sizes of Tubing at 1,000 psia

$$\gamma_g = 0.625$$
$$t = 130\ ^\circ F$$
$$p = 1{,}000\ psia$$

Tubing		Minimum Rate of Flow, MMcfd	
		Water	Condensate
Nominal	ID (in.)	u = 9.02 ft/sec	u = 5.81 ft/sec
2⅜	1.995	1.14	0.73
2⅞	2.441	1.71	1.10
3½	2.992	2.56	1.65
4	3.476	3.46	2.23
4½	4.082	4.77	3.07

TABLE 11–5.
Minimum Rates of Flow Required to Lift Water and Condensate in Various Sizes of Tubing at 200 psia

$$\gamma_g = 0.625$$
$$t = 130\ ^\circ F$$
$$p = 200\ psia$$

Tubing		Minimum Rate of Flow, MMcfd	
		Water	Condensate
Nominal	ID (in.)	u = 20.37 ft/sec	u = 13.18 ft/sec
1	1.049	0.13	0.08
1¼	1.380	0.22	0.15
1½	1.610	0.31	0.20
2⅜	1.995	0.47	0.30
2⅞	2.441	0.70	0.45
3½	2.992	1.06	0.68
4	3.476	1.42	0.92
4½	4.082	1.96	1.27

productive interval in a well, it is necessary to set the tubing near the bottom of the productive zone. A low-capacity well with 2⅜-in. tubing set several hundred feet above the productive zone in 5½-in. casing very likely will load up with liquids. Here, the minimum rate of flow would be determined by the velocity required to lift liquids in the 5½-in. casing.

Turning attention to the low-pressure gas wells now being completed in shallow producing areas, some of the problems in lifting liquids from low-pressure wells can be identified. At current sale prices, a well producing 100 Mcfd at a wellhead pressure of 200 psia produces sufficient income to justify operation. However, an examination of the results of calculations given in table 11–5 shows that water probably could not be lifted through 1-in. nominal tubing. If production is impeded by the accumulation of water in the wellbore, the only remedy remaining is to install a mechanical pump. Indeed, removing liquids with mechanical pumps from small-capacity gas wells is becoming a common method of operation.

All of the foregoing discussion is premised on the assumption that the well produces very moderate amounts of liquid. A large-enough quantity of water can kill almost any gas well no matter how large its capacity.

PROBLEMS

1. A small-capacity gas well with a depth of 600 ft and a shutin pressure at the bottom of the hole of 260 psia has an open-flow capacity of 750 Mcfd and an exponent of 0.85. Estimate the largest size of tubing and rate of flow that will still lift water at the initial bottom-hole pressure of 260 psia when the pressure has been reduced to 150 psia.

2. Calculate a table similar to tables 11–4 and 11–5 for a well operating at a pressure of 3,000 psia and 200°F at the wellhead.

REFERENCES

1. J.O. Duggan. "Estimating Flow Rate Required to Keep Gas Wells Unloaded." *J. Pet. Tech.* (December 1961), p. 1173.
2. R.G. Turner, M.G. Hubbard, and A.E. Dukler. "Analysis and Prediction of Minimum Flow Rate for the Continuous Removal of Liquids from Gas Wells." *J. Pet. Tech. Trans. AIME*, 246, (November 1969) 1175.

12 Gas Sales Contracts

A gas sales contract is a legal document between a seller who owns, controls, or has the right to sell and dispose of the gas to be produced from certain oil and gas leaseholds and a buyer who is engaged or is to be engaged in buying and selling gas in wholesale quantities. Usually, the buyer is engaged also in transporting gas in a pipeline system. As such, the gas sales contract is primarily the concern of attorneys and accountants, but the natural gas engineer makes important contributions in the successful administration of a gas sales contract. Therefore, it is not the intent of this chapter to outline a model gas sales contract but to bring out those areas of the contract that merit the attention of the natural gas engineer. Although gas sales contracts may vary in content, the contents have enough in common for the principal features to be discussed in general terms.

Lands or leaseholds dedicated. Contracts are usually specific as to the leaseholds dedicated to the performance of a certain contract. Such information determines the economic factors that are important in evaluating the repair or the redrilling of old wells or a proposal to drill new wells on dedicated acreage. If the acreage is not dedicated, the factors for determining the economics of a new well can be estimated on the basis of an anticipated contract.

Quantity and rate of take. The basic provisions of a contract are the quantity of gas to be sold, the rate at which the gas is to be taken, the quality of the gas, and the price. The provisions vary from contract to contract with the quantity of gas to be taken over a period of time, usually a year, set proportional to the mutually agreed recoverable reserve for the leases or to a specific statement of an average quantity of gas per day. The daily rate at which the gas is taken may vary from day to day, but the buyer may be required to take an average daily quantity of gas over a period of time, usually a year. Such clauses may vary from a daily, fixed rate to 67% of yearly quantity taken in six specific winter months with the remaining 33% of the yearly quantity taken during the six summer months. In case the seller's wells cannot meet the rate requirements of the contract, alternative rate requirements become controlling.

In areas or fields in which a regulatory body sets allowable production for the wells, the allowables become the controlling factors for rates of production. Under these conditions, the buyers usually have the responsibility for making nominations for gas takes. Contracts may specify how such nominations of quantities of gas are to be determined.

In any event, the engineer responsible for the performance of the wells should have detailed knowledge as to how the contractual or allowable rates of production are determined.

Quality of gas. Contracts are usually specific as to the quality of the gas. Rigid limits may be placed on dust, liquids, hydrogen sulfide and total sulfur contents. The seller may be required to dehydrate the gas so that the water vapor content may be no more than 7 lb/MMcf or some similar water vapor content. Also, a minimum heating value for the gas may be specified. If the heating value should fall below the minimum, the buyer may have the option of refusing to take the gas.

Quality requirements have a strong influence on the type and cost of surface equipment installed at the seller's expense at the well. Therefore, these requirements cannot be ignored by those responsible for approving cost estimates for well completions.

Price. Contracts contain various price provisions ranging from a fixed, stated price to an acceptance of the maximum lawful price established in section 109 of the Natural Gas Policy Act of 1978 (NGPA) computed in accordance with subsection (b) of section 109 of the NGPA. The contract may contain provisions for an increase in sales price, providing the gas becomes qualified under the NGPA. In determining future prices, the engineer is well advised to seek competent assistance.

Natural gas is sold either by volume, usually in units of 1,000 cu ft (Mcf) at measurement pressure and temperature bases or by the heating value in units of 1 million British thermal units (Btu). If sold by its heating value, the determination of the value becomes as important as the accuracy of the measuring.

Definitions. Most contracts set out the definitions for the terms under which the gas is sold. The definitions may include gross heating value, British thermal units, day, contract year, measurement bases, and other terms important to an understanding of the contract. It is possible that a contract might define the standard cubic foot of gas for measurement to be 14.65 psia and 60°F, the standard cubic foot for pricing purposes to be 14.73 psia and 60°F, and for the gross heating value the standard cubic foot to be at a pressure of 30 in. of mercury at 32°F and a temperature of 60°F. The only safe way through the thicket of definitions is through the sales contract.

Point of delivery. One of the most important pieces of information for the field engineer is the point of delivery because the seller is usually responsible for the equipment up to that point. The title to the gas passes from the seller to the buyer at the point of delivery. Also, the contract may state that the seller shall deliver gas at the varying pressures maintained at the delivery point, which means that the seller must furnish the compression facilities that may be necessary to deliver the gas.

Gas measurement. The gas measurement section of the contract usually states that the orifice meters shall be installed, operated, and gas volumes computed in accordance with the recommendations prescribed in "*Orifice Metering of Natural Gas,*" Gas Measurement Committee Report No. 3 dated April 1955 of the American Gas Association or such subsequent changes as are mutually agreed upon. The contract may specify the barometric pressure to be used in calculating absolute pressures and may adopt constant values for the Reynolds number factor, F_r, the expansion factor, Y, the supercompressibility factor, F_{pv}, and the manometer factor, F_m (see equation 4–5). Other factors may be specified unless they are measured.

Also, the gas measurement provisions may set specific times for calibrating the meter, checking the orifice plate, measuring specific gravity, determining heating value and setting procedures for adjusting errors in gas measurement. All of these are important considerations for the seller.

SUMMARY

The sale of natural gas differs from that of crude oil in several ways. Gas cannot be stored on the lease except by leaving it unproduced. Therefore, the seller must have his wells ready to produce when there is a market for the gas. The price paid for gas is subject to complicated regulation at present writing, and the price may vary from well to well even though the wells are in the same reservoir. As a result, the engineer involved in the management and operation of gas wells must have a working knowledge of the sales contracts that govern the sale of the gas. The pertinent features of the contracts that influence engineering decisions have been identified.

APPENDIX A. Compressibility Factors for Natural Gas

TABLE A–1.

*Pseudocritical Properties of Hydrocarbon Gases, p_{pc} and T_{pc} **

γ_g	p_{pc}	T_{pc}	γ_g	p_{pc}	T_{pc}
0.55	673	336	0.85	664	441
0.56	673	341	0.86	664	444
0.57	672	346	0.87	663	448
0.58	672	350	0.88	663	451
0.59	672	354	0.89	662	454
0.60	671	358	0.90	662	457
0.61	671	362	0.91	662	461
0.62	671	365	0.92	662	464
0.63	670	368	0.93	661	467
0.64	670	372	0.94	661	471
0.65	670	375	0.95	660	474
0.66	670	378	0.96	660	477
0.67	669	382	0.97	659	481
0.68	669	385	0.98	659	484
0.69	669	388	0.99	659	487
0.70	668	392	1.00	658	491
0.71	668	395	1.01	658	494
0.72	668	398	1.02	657	497
0.73	668	401	1.03	656	500
0.74	667	405	1.04	656	504
0.75	667	408	1.05	655	507
0.76	667	411	1.06	655	510
0.77	666	415	1.07	654	514
0.78	666	418	1.08	654	517
0.79	666	421	1.09	653	520
0.80	665	424	1.10	652	524
0.81	665	428	1.11	652	527
0.82	665	431	1.12	651	530
0.83	665	434	1.13	651	534
0.84	664	438	1.14	650	537

Do not interpolate; values are inclusive to the next higher value.

* Used by permission of the Pacific Energy Association (formerly the California Natural Gasoline Association), taken from Bulletin No. TS–461, 1947. Data are for the average of the natural gas mixtures occuring in the western states—primarily California.

TABLE A–2.

*Correction for Pseudocritical Properties of Hydrocarbon Gases for Carbon Dioxide, Nitrogen, and Hydrogen Sulfide**

Add if positive or subtract if negative from values in table A–1.

Volume Percent of CO_2 or N_2 in Gas	Carbon Dioxide, CO_2		Nitrogen, N_2		Hydrogen sulfide, H_2S	
	p_{pc}	T_{pc}	p_{pc}	T_{pc}	p_{pc}	T_{pc}
1	+ 4	− 1	− 1	− 3	+ 2	−2
2	+ 8	− 3	− 3	− 6	+ 4	−2
3	+ 12	− 5	− 5	− 9	+ 7	−3
4	+ 17	− 7	− 7	−11	+10	−3
5	+ 21	− 9	− 9	−14	+15	−4
6	+ 25	−11	−11	−17		
7	+ 30	−12	−12	−20		
8	+ 34	−14	−14	−22		
9	+ 39	−16	−16	−25		
10	+ 44	−17	−17	−28		
11	+ 48	−19	−19	−30		
12	+ 53	−21	−21	−33		
13	+ 57	−22	−22	−36		
14	+ 61	−24	−24	−39		
15	+ 66	−26	−26	−42		
16	+ 70	−27	−27	−44		
17	+ 74	−29	−29	−47		
18	+ 79	−31	−31	−50		
19	+ 83	−32	−32	−52		
20	+ 87	−34	−34	−55		
21	+ 92	−36	−36	−58		
22	+ 96	−37	−37	−60		
23	+100	−39	−39	−63		
24	+104	−41	−41	−66		
25	+109	−42	−42	−68		
26	+113	−44	−44	−71		
27	+117	−46	−46	−74		
28	+122	−47	−47	−77		
29	+126	−49	−49	−79		
30	+130	−51	−51	−82		
31	+134	−52	−52	−85		
32	+139	−54	−54	−87		
33	+143	−56	−56	−90		
34	+147	−57	−57	−93		
35	+152	−59	−59	−95		
36	+156	−61	−61	−98		

* Values for carbon dioxide and nitrogen are based on data from the California Natural Gasoline Association Bulletin No. TS–461. Used by permision of the Interstate Oil Compact Commission. Values for hydrogen sulfide are supplied by the author.

TABLE A–3.

Compressibility Factors for Natural Gases at various Pseudoreduced Pressures, p_{pr}, and Pseudoreduced Temperatures, T_{pr} *

Pseudo-reduced Pressure, p_{pr}	*Pseudoreduced Temperature,* T_{pr}									
	1.05	1.10	1.15	1.20	1.25	1.30	1.35	1.40	1.45	1.50
0.20	0.938	0.948	0.953	0.959	0.964	0.969	0.971	0.973	0.978	0.979
0.25	0.922	0.934	0.941	0.949	0.955	0.960	0.964	0.968	0.972	0.974
0.30	0.905	0.920	0.928	0.938	0.946	0.951	0.956	0.962	0.965	0.969
0.35	0.887	0.905	0.914	0.927	0.937	0.943	0.949	0.956	0.959	0.964
0.40	0.868	0.889	0.900	0.916	0.928	0.934	0.942	0.949	0.953	0.959
0.45	0.849	0.872	0.887	0.905	0.919	0.926	0.934	0.943	0.948	0.954
0.50	0.829	0.855	0.874	0.894	0.909	0.918	0.926	0.937	0.942	0.949
0.55	0.810	0.839	0.862	0.883	0.900	0.909	0.919	0.930	0.937	0.944
0.60	0.790	0.822	0.849	0.872	0.890	0.900	0.912	0.923	0.931	0.939
0.65	0.769	0.804	0.835	0.861	0.881	0.893	0.905	0.917	0.926	0.935
0.70	0.747	0.785	0.820	0.850	0.871	0.885	0.898	0.911	0.921	0.930
0.75	0.722	0.767	0.806	0.839	0.861	0.878	0.890	0.905	0.916	0.925
0.80	0.698	0.748	0.791	0.827	0.851	0.870	0.883	0.899	0.910	0.920
0.85	0.672	0.729	0.776	0.815	0.842	0.861	0.876	0.893	0.905	0.916
0.90	0.645	0.710	0.761	0.802	0.832	0.852	0.870	0.887	0.899	0.911

* Table A–3, in its entirety, was taken from F.H. Poettmann and P.G. Carpenter, "The Multiphase Flow of Gas, Oil, and Water through Vertical Flow Strings with Application to the Design of Gas-Lift Installations," *API Drilling and Production Practice*, pp. 279–91 (1952). Use herein was granted by permission of the American Petroleum Institute.

Table A–3, continued

Pseudo-reduced Pressure, p_{pr}	*Pseudoreduced Temperature,* T_{pr}									
	1.05	1.10	1.15	1.20	1.25	1.30	1.35	1.40	1.45	1.50
0.95	0.618	0.690	0.748	0.791	0.823	0.845	0.863	0.881	0.895	0.906
1.00	0.590	0.670	0.734	0.780	0.813	0.838	0.856	0.875	0.890	0.901
1.05	0.555	0.647	0.719	0.768	0.803	0.829	0.849	0.869	0.885	0.897
1.10	0.520	0.624	0.703	0.755	0.793	0.820	0.842	0.862	0.879	0.893
1.15	0.482	0.602	0.688	0.743	0.783	0.812	0.836	0.856	0.874	0.889
1.20	0.444	0.580	0.672	0.731	0.773	0.804	0.829	0.850	0.869	0.884
1.25	0.398	0.555	0.656	0.719	0.763	0.797	0.823	0.845	0.864	0.880
1.30	0.351	0.530	0.640	0.707	0.753	0.789	0.817	0.839	0.859	0.876
1.35	0.307	0.505	0.623	0.695	0.743	0.781	0.810	0.834	0.855	0.872
1.40	0.263	0.480	0.605	0.682	0.732	0.772	0.803	0.828	0.850	0.868
1.45	0.259	0.453	0.588	0.670	0.721	0.765	0.797	0.824	0.846	0.864
1.50	0.254	0.425	0.570	0.657	0.710	0.757	0.791	0.817	0.841	0.859
1.55	0.253	0.409	0.556	0.646	0.701	0.750	0.785	0.811	0.837	0.855
1.60	0.251	0.393	0.541	0.634	0.691	0.742	0.779	0.805	0.832	0.851
1.65	0.251	0.386	0.529	0.623	0.682	0.735	0.773	0.800	0.828	0.847
1.70	0.252	0.378	0.517	0.612	0.672	0.728	0.767	0.795	0.823	0.843
1.75	0.256	0.374	0.507	0.602	0.664	0.720	0.761	0.790	0.819	0.840

1.80	0.260	0.370	0.497	0.592	0.656	0.712	0.755	0.784	0.815	0.837
1.85	0.265	0.370	0.489	0.582	0.648	0.706	0.750	0.779	0.811	0.834
1.90	0.270	0.369	0.480	0.572	0.640	0.699	0.744	0.774	0.806	0.830
1.95	0.275	0.369	0.474	0.564	0.634	0.693	0.738	0.771	0.803	0.827
2.00	0.280	0.370	0.467	0.555	0.627	0.687	0.732	0.767	0.799	0.824
2.05	0.286	0.371	0.462	0.547	0.620	0.681	0.727	0.762	0.795	0.821
2.10	0.292	0.372	0.457	0.539	0.612	0.674	0.721	0.757	0.791	0.818
2.15	0.298	0.374	0.454	0.534	0.607	0.669	0.716	0.753	0.788	0.815
2.20	0.305	0.376	0.450	0.528	0.602	0.663	0.711	0.748	0.784	0.811
2.25	0.311	0.378	0.449	0.525	0.598	0.658	0.707	0.745	0.781	0.808
2.30	0.318	0.380	0.448	0.522	0.593	0.652	0.702	0.741	0.778	0.805
2.35	0.324	0.383	0.448	0.521	0.590	0.649	0.699	0.737	0.775	0.803
2.40	0.330	0.387	0.449	0.520	0.586	0.645	0.695	0.733	0.771	0.800
2.45	0.336	0.390	0.450	0.520	0.583	0.642	0.692	0.731	0.768	0.798
2.50	0.343	0.394	0.451	0.519	0.581	0.638	0.688	0.728	0.765	0.796
2.55	0.349	0.398	0.454	0.519	0.581	0.636	0.685	0.725	0.763	0.793
2.60	0.356	0.402	0.458	0.520	0.580	0.633	0.682	0.722	0.760	0.791
2.65	0.362	0.406	0.460	0.521	0.580	0.631	0.680	0.720	0.758	0.789
2.70	0.369	0.410	0.463	0.522	0.579	0.629	0.678	0.718	0.756	0.787
2.75	0.375	0.414	0.467	0.524	0.579	0.628	0.676	0.716	0.754	0.785
2.80	0.381	0.419	0.472	0.526	0.579	0.627	0.673	0.713	0.752	0.783
2.85	0.387	0.424	0.476	0.528	0.579	0.626	0.672	0.712	0.750	0.781
2.90	0.394	0.429	0.481	0.530	0.580	0.625	0.670	0.710	0.748	0.779

Table A–3, continued

Pseudo-reduced Pressure, p_{pr}	*Pseudoreduced Temperature,* T_{pr}									
	1.60	1.70	1.80	1.90	2.00	2.20	2.40	2.60	2.80	3.00
0.20	0.984	0.987	0.990	0.991	0.993	0.995	0.998	0.999	1.000	1.001
0.25	0.980	0.984	0.988	0.990	0.992	0.995	0.998	0.999	1.000	1.001
0.30	0.976	0.981	0.985	0.988	0.990	0.994	0.997	0.999	1.000	1.001
0.35	0.972	0.978	0.983	0.986	0.989	0.994	0.997	0.999	1.000	1.001
0.40	0.968	0.975	0.980	0.983	0.987	0.992	0.996	0.998	1.000	1.002
0.45	0.964	0.972	0.978	0.981	0.985	0.991	0.995	0.998	1.000	1.002
0.50	0.960	0.969	0.976	0.979	0.983	0.990	0.994	0.998	1.000	1.002
0.55	0.956	0.966	0.974	0.978	0.982	0.989	0.994	0.998	1.000	1.002
0.60	0.952	0.963	0.971	0.976	0.981	0.988	0.993	0.997	1.000	1.003
0.65	0.949	0.961	0.969	0.974	0.980	0.987	0.993	0.997	1.000	1.003
0.70	0.945	0.958	0.967	0.972	0.978	0.986	0.992	0.997	1.001	1.004
0.75	0.942	0.955	0.965	0.971	0.977	0.985	0.992	0.997	1.001	1.004
0.80	0.938	0.952	0.962	0.969	0.975	0.984	0.991	0.997	1.001	1.005
0.85	0.935	0.950	0.960	0.967	0.974	0.983	0.991	0.997	1.001	1.005
0.90	0.931	0.947	0.958	0.965	0.972	0.983	0.990	0.996	1.001	1.006
0.95	0.927	0.944	0.956	0.963	0.971	0.982	0.990	0.996	1.001	1.006
1.00	0.923	0.941	0.953	0.961	0.970	0.981	0.989	0.996	1.001	1.007
1.05	0.920	0.939	0.951	0.960	0.969	0.980	0.989	0.996	1.001	1.007
1.10	0.917	0.936	0.949	0.958	0.967	0.979	0.988	0.996	1.002	1.008
1.15	0.913	0.933	0.947	0.957	0.966	0.979	0.988	0.996	1.002	1.008
1.20	0.909	0.930	0.945	0.955	0.964	0.978	0.987	0.996	1.002	1.008
1.25	0.905	0.928	0.943	0.954	0.963	0.978	0.987	0.996	1.002	1.008
1.30	0.901	0.925	0.941	0.952	0.962	0.977	0.987	0.996	1.002	1.009
1.35	0.898	0.923	0.940	0.951	0.961	0.976	0.986	0.996	1.002	1.009
1.40	0.895	0.920	0.938	0.949	0.960	0.975	0.986	0.996	1.002	1.010

1.45	0.892	0.918	0.936	0.947	0.959	0.975	0.986	0.996	1.002	1.010
1.50	0.889	0.915	0.934	0.946	0.957	0.974	0.985	0.995	1.003	1.010
1.55	0.886	0.913	0.932	0.945	0.956	0.973	0.985	0.995	1.003	1.010
1.60	0.882	0.910	0.930	0.943	0.954	0.972	0.984	0.995	1.003	1.011
1.65	0.880	0.908	0.929	0.942	0.953	0.972	0.984	0.995	1.003	1.011
1.70	0.877	0.905	0.927	0.941	0.952	0.971	0.983	0.995	1.004	1.012
1.75	0.874	0.903	0.925	0.940	0.951	0.971	0.983	0.995	1.004	1.012
1.80	0.871	0.900	0.923	0.938	0.950	0.970	0.983	0.995	1.004	1.012
1.85	0.869	0.899	0.922	0.937	0.949	0.970	0.983	0.995	1.004	1.012
1.90	0.866	0.897	0.921	0.936	0.948	0.969	0.982	0.995	1.005	1.013
1.95	0.864	0.895	0.920	0.935	0.948	0.969	0.982	0.995	1.005	1.013
2.00	0.861	0.893	0.918	0.933	0.947	0.968	0.982	0.995	1.006	1.014
2.05	0.859	0.891	0.917	0.932	0.946	0.968	0.982	0.995	1.006	1.014
2.10	0.856	0.889	0.915	0.931	0.945	0.967	0.981	0.995	1.007	1.015
2.15	0.854	0.887	0.914	0.930	0.945	0.967	0.981	0.995	1.007	1.015
2.20	0.851	0.885	0.912	0.929	0.944	0.966	0.981	0.995	1.007	1.016
2.25	0.850	0.884	0.911	0.928	0.943	0.966	0.981	0.995	1.007	1.016
2.30	0.848	0.882	0.909	0.927	0.942	0.965	0.980	0.995	1.008	1.017
2.35	0.846	0.881	0.908	0.927	0.942	0.965	0.980	0.995	1.008	1.017
2.40	0.843	0.879	0.907	0.926	0.941	0.964	0.980	0.995	1.008	1.018
2.45	0.842	0.878	0.906	0.925	0.941	0.964	0.980	0.995	1.008	1.018
2.50	0.840	0.876	0.904	0.924	0.941	0.963	0.980	0.995	1.009	1.019
2.55	0.839	0.875	0.903	0.923	0.941	0.963	0.980	0.995	1.009	1.019
2.60	0.837	0.873	0.902	0.922	0.940	0.963	0.980	0.995	1.009	1.020
2.65	0.835	0.872	0.901	0.922	0.940	0.963	0.980	0.995	1.009	1.020
2.70	0.833	0.871	0.900	0.921	0.939	0.962	0.980	0.996	1.010	1.021
2.75	0.832	0.870	0.900	0.921	0.939	0.962	0.980	0.996	1.010	1.021
2.80	0.830	0.869	0.899	0.920	0.938	0.962	0.980	0.966	1.010	1.022
2.85	0.829	0.868	0.898	0.920	0.938	0.962	0.980	0.996	1.010	1.022
2.90	0.828	0.867	0.897	0.919	0.938	0.962	0.980	0.997	1.011	1.023

Table A–3, continued

Pseudo-reduced Pressure, p_{pr}	*Pseudoreduced Temperature,* T_{pr}									
	1.05	1.10	1.15	1.20	1.25	1.30	1.35	1.40	1.45	1.50
2.95	0.401	0.434	0.485	0.532	0.580	0.625	0.670	0.709	0.747	0.778
3.00	0.407	0.440	0.489	0.534	0.581	0.624	0.669	0.707	0.745	0.777
3.05	0.413	0.446	0.494	0.537	0.582	0.624	0.669	0.706	0.744	0.776
3.10	0.420	0.452	0.499	0.540	0.584	0.625	0.668	0.705	0.742	0.775
3.15	0.426	0.457	0.503	0.543	0.586	0.625	0.668	0.704	0.742	0.774
3.20	0.432	0.463	0.507	0.546	0.588	0.626	0.668	0.703	0.741	0.773
3.25	0.438	0.469	0.511	0.548	0.590	0.627	0.668	0.701	0.741	0.773
3.30	0.445	0.475	0.516	0.551	0.592	0.628	0.669	0.702	0.740	0.772
3.35	0.451	0.481	0.520	0.555	0.594	0.629	0.669	0.702	0.740	0.772
3.40	0.458	0.488	0.525	0.559	0.597	0.631	0.670	0.703	0.739	0.771
3.45	0.464	0.494	0.530	0.562	0.599	0.632	0.670	0.703	0.739	0.771
3.50	0.471	0.500	0.535	0.566	0.602	0.633	0.670	0.704	0.739	0.771
3.55	0.477	0.505	0.539	0.569	0.605	0.635	0.672	0.705	0.739	0.771
3.60	0.483	0.511	0.544	0.572	0.608	0.637	0.673	0.706	0.740	0.771
3.65	0.490	0.517	0.548	0.577	0.611	0.639	0.675	0.707	0.740	0.771
3.70	0.497	0.523	0.553	0.582	0.614	0.641	0.677	0.708	0.741	0.772
3.75	0.503	0.529	0.558	0.586	0.617	0.642	0.678	0.709	0.741	0.772
3.80	0.509	0.535	0.563	0.590	0.620	0.644	0.680	0.710	0.742	0.773
3.85	0.515	0.541	0.568	0.594	0.623	0.646	0.681	0.711	0.742	0.773
3.90	0.522	0.547	0.573	0.599	0.627	0.649	0.683	0.713	0.743	0.774
3.95	0.528	0.552	0.577	0.603	0.630	0.651	0.685	0.714	0.744	0.775
4.00	0.534	0.558	0.582	0.608	0.633	0.653	0.687	0.716	0.746	0.776
4.05	0.540	0.564	0.587	0.612	0.636	0.656	0.689	0.717	0.747	0.777
4.10	0.547	0.570	0.592	0.616	0.640	0.659	0.692	0.719	0.749	0.779
4.15	0.553	0.575	0.596	0.620	0.644	0.662	0.695	0.720	0.750	0.780

4.20	0.560	0.581	0.601	0.624	0.648	0.665	0.697	0.722	0.752	0.781
4.25	0.566	0.587	0.606	0.628	0.651	0.668	0.699	0.724	0.753	0.782
4.30	0.572	0.593	0.611	0.632	0.655	0.671	0.701	0.726	0.755	0.784
4.35	0.579	0.598	0.616	0.636	0.658	0.674	0.703	0.728	0.757	0.785
4.40	0.586	0.604	0.621	0.641	0.662	0.678	0.706	0.730	0.759	0.787
4.45	0.592	0.610	0.626	0.645	0.666	0.680	0.709	0.732	0.761	0.788
4.50	0.599	0.616	0.631	0.650	0.670	0.683	0.712	0.734	0.763	0.790
4.55	0.605	0.621	0.636	0.654	0.673	0.687	0.715	0.736	0.765	0.792
4.60	0.612	0.627	0.641	0.659	0.677	0.691	0.718	0.739	0.768	0.794
4.65	0.617	0.633	0.646	0.663	0.681	0.694	0.720	0.741	0.770	0.796
4.70	0.623	0.639	0.651	0.668	0.685	0.698	0.723	0.743	0.772	0.798
4.75	0.630	0.644	0.655	0.672	0.689	0.701	0.726	0.746	0.774	0.800
4.80	0.637	0.650	0.662	0.677	0.693	0.705	0.729	0.749	0.777	0.802
4.85	0.643	0.655	0.667	0.681	0.697	0.708	0.732	0.751	0.779	0.804
4.90	0.650	0.661	0.672	0.686	0.701	0.712	0.736	0.753	0.782	0.806
4.95	0.656	0.666	0.677	0.690	0.705	0.716	0.739	0.756	0.784	0.808
5.00	0.663	0.672	0.683	0.694	0.709	0.720	0.742	0.760	0.787	0.810
5.05	0.669	0.677	0.688	0.698	0.713	0.723	0.745	0.763	0.789	0.812
5.10	0.675	0.683	0.693	0.703	0.717	0.727	0.749	0.766	0.792	0.814
5.15	0.681	0.689	0.698	0.708	0.721	0.731	0.752	0.769	0.795	0.817
5.20	0.688	0.695	0.703	0.713	0.725	0.735	0.755	0.772	0.798	0.820
5.25	0.694	0.700	0.708	0.717	0.729	0.739	0.758	0.775	0.800	0.822
5.30	0.701	0.706	0.713	0.722	0.734	0.743	0.762	0.778	0.803	0.825
5.35	0.707	0.712	0.718	0.726	0.738	0.747	0.765	0.781	0.806	0.827
5.40	0.713	0.718	0.723	0.731	0.742	0.751	0.769	0.785	0.809	0.830
5.45	0.720	0.723	0.728	0.735	0.746	0.755	0.772	0.788	0.812	0.832
5.50	0.727	0.729	0.733	0.740	0.751	0.759	0.776	0.792	0.815	0.835
5.55	0.733	0.734	0.738	0.745	0.755	0.763	0.779	0.795	0.818	0.837
5.60	0.739	0.740	0.744	0.750	0.760	0.768	0.783	0.799	0.821	0.840
5.65	0.745	0.745	0.749	0.755	0.764	0.772	0.787	0.802	0.824	0.843

Table A–3, continued

Pseudo-reduced Pressure, p_{pr}	*Pseudoreduced Temperature,* T_{pr}									
	1.60	1.70	1.80	1.90	2.00	2.20	2.40	2.60	2.80	3.00
2.95	0.827	0.866	0.897	0.919	0.938	0.962	0.980	0.997	1.011	1.023
3.00	0.825	0.864	0.896	0.918	0.938	0.962	0.981	0.997	1.012	1.024
3.05	0.823	0.863	0.896	0.918	0.938	0.962	0.981	0.997	1.012	1.024
3.10	0.822	0.862	0.895	0.917	0.937	0.962	0.981	0.998	1.013	1.025
3.15	0.822	0.862	0.895	0.917	0.937	0.962	0.981	0.998	1.013	1.025
3.20	0.821	0.861	0.894	0.917	0.937	0.962	0.982	0.998	1.014	1.026
3.25	0.820	0.860	0.894	0.917	0.937	0.962	0.982	0.998	1.014	1.026
3.30	0.819	0.859	0.893	0.916	0.937	0.962	0.982	0.999	1.015	1.027
3.35	0.819	0.859	0.893	0.916	0.937	0.962	0.982	0.999	1.015	1.027
3.40	0.818	0.858	0.892	0.916	0.937	0.963	0.983	1.000	1.016	1.028
3.45	0.818	0.858	0.892	0.916	0.937	0.963	0.983	1.000	1.016	1.029
3.50	0.817	0.858	0.891	0.915	0.937	0.963	0.984	1.000	1.017	1.030
3.55	0.817	0.858	0.891	0.915	0.937	0.963	0.984	1.000	1.017	1.030
3.60	0.816	0.857	0.890	0.915	0.937	0.964	0.985	1.001	1.018	1.031
3.65	0.816	0.857	0.890	0.915	0.937	0.964	0.985	1.001	1.019	1.031
3.70	0.816	0.856	0.890	0.916	0.937	0.965	0.986	1.002	1.019	1.032
3.75	0.816	0.856	0.890	0.916	0.937	0.965	0.986	1.002	1.019	1.032
3.80	0.816	0.855	0.891	0.916	0.937	0.966	0.987	1.003	1.020	1.033
3.85	0.816	0.855	0.891	0.916	0.937	0.966	0.988	1.004	1.021	1.033
3.90	0.817	0.855	0.891	0.917	0.938	0.967	0.989	1.005	1.022	1.034
3.95	0.817	0.855	0.891	0.917	0.938	0.967	0.989	1.006	1.022	1.034
4.00	0.818	0.856	0.892	0.917	0.939	0.968	0.990	1.007	1.023	1.035
4.05	0.819	0.856	0.892	0.917	0.939	0.968	0.991	1.008	1.023	1.035
4.10	0.820	0.857	0.893	0.918	0.940	0.969	0.992	1.009	1.024	1.036
4.15	0.821	0.857	0.894	0.918	0.940	0.970	0.993	1.010	1.025	1.036

4.20	0.822	0.859	0.895	0.919	0.941	0.971	0.994	1.011	1.026	1.037
4.25	0.823	0.859	0.896	0.919	0.941	0.972	0.995	1.012	1.026	1.037
4.30	0.824	0.861	0.897	0.920	0.942	0.973	0.996	1.013	1.027	1.038
4.35	0.825	0.861	0.897	0.921	0.942	0.973	0.997	1.013	1.028	1.039
4.40	0.826	0.862	0.898	0.922	0.943	0.974	0.998	1.014	1.029	1.040
4.45	0.827	0.862	0.899	0.922	0.943	0.975	0.999	1.015	1.029	1.040
4.50	0.829	0.864	0.900	0.923	0.944	0.977	1.000	1.016	1.030	1.041
4.55	0.830	0.865	0.901	0.924	0.945	0.977	1.001	1.017	1.031	1.042
4.60	0.832	0.867	0.902	0.925	0.946	0.978	1.002	1.018	1.032	1.043
4.65	0.833	0.868	0.903	0.926	0.947	0.979	1.003	1.019	1.033	1.043
4.70	0.835	0.869	0.904	0.927	0.948	0.981	1.004	1.020	1.034	1.044
4.75	0.836	0.870	0.905	0.928	0.949	0.982	1.005	1.021	1.035	1.045
4.80	0.838	0.872	0.907	0.929	0.950	0.983	1.007	1.022	1.036	1.046
4.85	0.840	0.873	0.908	0.930	0.951	0.984	1.008	1.023	1.037	1.047
4.90	0.842	0.875	0.910	0.932	0.952	0.985	1.009	1.024	1.038	1.048
4.95	0.844	0.877	0.911	0.933	0.953	0.986	1.010	1.025	1.038	1.048
5.00	0.846	0.879	0.912	0.934	0.954	0.988	1.011	1.027	1.039	1.049
5.05	0.848	0.880	0.914	0.935	0.955	0.989	1.012	1.028	1.040	1.050
5.10	0.850	0.882	0.916	0.937	0.957	0.990	1.013	1.029	1.041	1.051
5.15	0.851	0.884	0.917	0.938	0.958	0.991	1.014	1.030	1.042	1.051
5.20	0.853	0.886	0.919	0.940	0.960	0.992	1.016	1.031	1.043	1.052
5.25	0.855	0.887	0.920	0.941	0.961	0.993	1.017	1.032	1.044	1.052
5.30	0.858	0.889	0.921	0.943	0.963	0.995	1.018	1.033	1.045	1.053
5.35	0.860	0.891	0.923	0.944	0.964	0.996	1.019	1.034	1.046	1.053
5.40	0.862	0.893	0.925	0.946	0.966	0.998	1.021	1.035	1.047	1.055
5.45	0.865	0.895	0.926	0.947	0.967	0.999	1.022	1.036	1.048	1.056
5.50	0.868	0.897	0.928	0.949	0.969	1.000	1.023	1.038	1.049	1.057
5.55	0.870	0.899	0.929	0.950	0.970	1.001	1.024	1.039	1.050	1.058
5.60	0.872	0.900	0.931	0.952	0.972	1.003	1.026	1.040	1.051	1.058
5.65	0.874	0.902	0.933	0.953	0.974	1.004	1.027	1.041	1.052	1.059

Table A–3, continued

Pseudo-reduced Pressure, p_{pr}	*Pseudoreduced Temperature,* T_{pr}									
	1.05	1.10	1.15	1.20	1.25	1.30	1.35	1.40	1.45	1.50
5.70	0.751	0.751	0.754	0.760	0.769	0.777	0.791	0.806	0.827	0.846
5.75	0.757	0.756	0.759	0.764	0.773	0.781	0.795	0.809	0.830	0.848
5.80	0.763	0.762	0.764	0.769	0.778	0.785	0.799	0.813	0.833	0.851
5.85	0.769	0.768	0.770	0.774	0.782	0.789	0.803	0.817	0.836	0.854
5.90	0.775	0.774	0.777	0.779	0.787	0.794	0.807	0.821	0.839	0.857
5.95	0.781	0.779	0.782	0.784	0.792	0.798	0.810	0.824	0.842	0.860
6.00	0.788	0.785	0.787	0.790	0.797	0.803	0.814	0.828	0.845	0.863
6.05	0.794	0.790	0.792	0.795	0.801	0.807	0.818	0.831	0.848	0.865
6.10	0.800	0.796	0.797	0.800	0.805	0.812	0.822	0.835	0.852	0.868
6.15	0.805	0.801	0.802	0.805	0.809	0.816	0.826	0.838	0.855	0.870
6.20	0.811	0.807	0.808	0.810	0.814	0.820	0.830	0.842	0.858	0.873
6.25	0.816	0.812	0.813	0.815	0.818	0.824	0.834	0.846	0.861	0.876
6.30	0.822	0.818	0.818	0.820	0.823	0.829	0.839	0.850	0.864	0.880
6.35	0.829	0.824	0.823	0.825	0.828	0.833	0.843	0.853	0.867	0.883
6.40	0.835	0.830	0.829	0.830	0.833	0.838	0.848	0.857	0.871	0.886
6.45	0.841	0.835	0.834	0.835	0.837	0.842	0.852	0.861	0.874	0.889
6.50	0.848	0.841	0.839	0.840	0.842	0.847	0.856	0.865	0.878	0.892
6.55	0.853	0.846	0.844	0.845	0.846	0.851	0.860	0.868	0.881	0.895
6.60	0.858	0.852	0.850	0.851	0.851	0.855	0.864	0.871	0.884	0.898
6.65	0.864	0.857	0.855	0.856	0.856	0.859	0.868	0.875	0.888	0.900
6.70	0.870	0.863	0.860	0.861	0.861	0.863	0.872	0.880	0.892	0.903
6.75	0.875	0.868	0.865	0.866	0.865	0.867	0.876	0.884	0.895	0.906
6.80	0.881	0.874	0.871	0.872	0.870	0.872	0.880	0.888	0.899	0.910
6.85	0.886	0.879	0.876	0.877	0.875	0.877	0.884	0.891	0.902	0.913
6.90	0.892	0.885	0.881	0.881	0.881	0.881	0.888	0.895	0.906	0.917

6.95	0.897	0.890	0.886	0.885	0.885	0.885	0.892	0.898	0.909	0.919
7.00	0.903	0.896	0.892	0.890	0.890	0.890	0.896	0.902	0.913	0.922
7.05	0.909	0.901	0.897	0.895	0.894	0.897	0.899	0.906	0.916	0.925
7.10	0.915	0.907	0.903	0.900	0.899	0.899	0.903	0.910	0.920	0.929
7.15	0.921	0.912	0.908	0.905	0.903	0.903	0.907	0.914	0.923	0.932
7.20	0.928	0.918	0.913	0.910	0.908	0.907	0.911	0.918	0.927	0.935
7.25	0.933	0.923	0.918	0.915	0.912	0.910	0.915	0.921	0.930	0.938
7.30	0.938	0.929	0.923	0.920	0.917	0.914	0.919	0.925	0.934	0.942
7.35	0.943	0.934	0.928	0.924	0.921	0.918	0.923	0.928	0.937	0.946
7.40	0.949	0.940	0.933	0.929	0.926	0.923	0.927	0.932	0.941	0.950
7.45	0.955	0.945	0.938	0.934	0.930	0.927	0.931	0.936	0.944	0.953
7.50	0.961	0.951	0.943	0.939	0.935	0.932	0.935	0.940	0.948	0.956
7.55	0.966	0.956	0.948	0.944	0.939	0.936	0.939	0.944	0.951	0.959
7.60	0.972	0.961	0.953	0.949	0.944	0.941	0.943	0.949	0.955	0.962
7.65	0.987	0.966	0.958	0.954	0.949	0.945	0.947	0.952	0.958	0.965
7.70	0.983	0.972	0.963	0.959	0.954	0.950	0.952	0.955	0.962	0.969
7.75	0.989	0.977	0.968	0.963	0.958	0.954	0.956	0.959	0.965	0.972
7.80	0.995	0.983	0.973	0.968	0.962	0.959	0.960	0.963	0.969	0.976
7.85	1.000	0.988	0.978	0.973	0.966	0.963	0.964	0.967	0.973	0.979
7.90	1.005	0.993	0.984	0.978	0.970	0.968	0.968	0.972	0.977	0.982
7.95	1.011	0.999	0.989	0.983	0.974	0.972	0.972	0.976	0.980	0.986
8.00	1.017	1.005	0.995	0.988	0.979	0.977	0.976	0.980	0.984	0.990
8.05	1.022	1.010	1.000	0.993	0.983	0.982	0.980	0.984	0.988	0.994
8.10	1.027	1.016	1.006	0.998	0.987	0.987	0.984	0.989	0.992	0.998
8.15	1.032	1.021	1.011	1.002	0.991	0.991	0.988	0.992	0.996	1.000
8.20	1.038	1.026	1.016	1.007	0.996	0.996	0.992	0.996	1.000	1.003
8.25	1.044	1.031	1.021	1.012	1.000	1.000	0.996	0.999	1.004	1.007
8.30	1.050	1.037	1.026	1.017	1.005	1.004	1.001	1.003	1.008	1.011
8.35	1.055	1.042	1.031	1.021	1.010	1.008	1.005	1.007	1.011	1.014
8.40	1.060	1.048	1.036	1.026	1.015	1.012	1.010	1.012	1.015	1.018

Table A–3, continued

Pseudo-reduced Pressure, p_{pr}	*Pseudoreduced Temperature,* T_{pr}									
	1.60	1.70	1.80	1.90	2.00	2.20	2.40	2.60	2.80	3.00
5.70	0.877	0.905	0.935	0.955	0.976	1.006	1.028	1.042	1.053	1.060
5.75	0.879	0.907	0.937	0.957	0.977	1.007	1.029	1.043	1.054	1.061
5.80	0.882	0.909	0.939	0.959	0.979	1.008	1.031	1.045	1.055	1.062
5.85	0.884	0.911	0.940	0.961	0.980	1.009	1.032	1.046	1.056	1.063
5.90	0.887	0.913	0.942	0.963	0.982	1.011	1.033	1.048	1.057	1.064
5.95	0.889	0.915	0.944	0.965	0.984	1.012	1.034	1.049	1.058	1.065
6.00	0.892	0.918	0.946	0.967	0.986	1.013	1.036	1.050	1.059	1.066
6.05	0.894	0.920	0.948	0.968	0.987	1.015	1.037	1.051	1.060	1.067
6.10	0.897	0.922	0.950	0.970	0.989	1.017	1.038	1.052	1.061	1.068
6.15	0.899	0.924	0.952	0.971	0.991	1.018	1.039	1.053	1.062	1.069
6.20	0.902	0.927	0.954	0.973	0.993	1.019	1.041	1.055	1.063	1.070
6.25	0.904	0.930	0.956	0.975	0.995	1.020	1.042	1.056	1.064	1.071
6.30	0.907	0.931	0.958	0.977	0.997	1.022	1.043	1.057	1.065	1.072
6.35	0.909	0.932	0.960	0.979	0.998	1.023	1.045	1.058	1.066	1.072
6.40	0.912	0.936	0.962	0.981	1.000	1.025	1.047	1.059	1.067	1.073
6.45	0.915	0.938	0.964	0.983	1.001	1.026	1.048	1.060	1.068	1.074
6.50	0.918	0.941	0.966	0.985	1.003	1.028	1.049	1.062	1.069	1.075
6.55	0.920	0.943	0.968	0.987	1.005	1.029	1.050	1.063	1.070	1.076
6.60	0.923	0.945	0.970	0.989	1.007	1.031	1.052	1.064	1.072	1.077
6.65	0.925	0.947	0.972	0.991	1.008	1.032	1.053	1.065	1.073	1.078
6.70	0.928	0.950	0.974	0.993	1.010	1.034	1.054	1.067	1.074	1.080
6.75	0.930	0.952	0.976	0.995	1.011	1.036	1.055	1.068	1.075	1.081
6.80	0.933	0.955	0.979	0.997	1.013	1.038	1.057	1.069	1.077	1.082
6.85	0.936	0.957	0.981	0.998	1.015	1.040	1.058	1.070	1.078	1.083
6.90	0.939	0.960	0.983	1.000	1.017	1.041	1.059	1.071	1.079	1.084

6.95	0.941	0.962	0.985	1.002	1.018	1.042	1.060	1.072	1.080	1.085
7.00	0.944	0.965	0.988	1.004	1.020	1.044	1.062	1.073	1.081	1.086
7.05	0.947	0.967	0.990	1.006	1.022	1.045	1.063	1.075	1.082	1.087
7.10	0.950	0.970	0.992	1.008	1.024	1.047	1.065	1.077	1.083	1.088
7.15	0.953	0.972	0.994	1.010	1.026	1.048	1.066	1.078	1.084	1.089
7.20	0.956	0.975	0.997	1.012	1.028	1.050	1.067	1.079	1.086	1.091
7.25	0.959	0.977	0.999	1.014	1.029	1.051	1.068	1.080	1.087	1.092
7.30	0.962	0.980	1.001	1.017	1.031	1.053	1.070	1.081	1.088	1.093
7.35	0.965	0.982	1.003	1.019	1.033	1.055	1.071	1.082	1.089	1.094
7.40	0.968	0.985	1.005	1.021	1.035	1.057	1.073	1.084	1.091	1.095
7.45	0.970	0.988	1.007	1.023	1.037	1.058	1.074	1.085	1.092	1.096
7.50	0.973	0.991	1.010	1.025	1.039	1.060	1.076	1.087	1.093	1.098
7.55	0.976	0.993	1.012	1.027	1.041	1.061	1.077	1.088	1.094	1.099
7.60	0.979	0.996	1.015	1.029	1.043	1.063	1.079	1.089	1.096	1.100
7.65	0.981	0.998	1.017	1.031	1.045	1.065	1.080	1.090	1.097	1.101
7.70	0.984	1.001	1.019	1.033	1.047	1.067	1.082	1.092	1.098	1.102
7.75	0.987	1.003	1.021	1.035	1.049	1.068	1.083	1.093	1.099	1.103
7.80	0.990	1.006	1.024	1.038	1.051	1.070	1.084	1.095	1.100	1.104
7.85	0.993	1.009	1.026	1.040	1.053	1.071	1.085	1.096	1.101	1.105
7.90	0.997	1.012	1.029	1.042	1.055	1.073	1.087	1.098	1.103	1.107
7.95	0.999	1.015	1.031	1.044	1.057	1.075	1.088	1.099	1.104	1.108
8.00	1.001	1.018	1.033	1.047	1.059	1.078	1.090	1.100	1.106	1.110
8.05	1.004	1.020	1.035	1.049	1.060	1.079	1.091	1.101	1.107	1.111
8.10	1.008	1.022	1.038	1.051	1.062	1.081	1.093	1.102	1.109	1.112
8.15	1.011	1.025	1.040	1.053	1.064	1.082	1.095	1.103	1.110	1.113
8.20	1.014	1.028	1.043	1.055	1.066	1.084	1.097	1.105	1.111	1.115
8.25	1.017	1.030	1.045	1.057	1.068	1.086	1.098	1.106	1.112	1.116
8.30	1.021	1.033	1.048	1.060	1.070	1.088	1.100	1.108	1.113	1.118
8.35	1.024	1.036	1.050	1.062	1.072	1.089	1.101	1.109	1.115	1.119
8.40	1.027	1.039	1.052	1.064	1.074	1.091	1.102	1.111	1.117	1.120

Table A–3, continued

Pseudo-reduced Pressure, p_{pr}	Pseudoreduced Temperature, T_{pr}									
	1.05	1.10	1.15	1.20	1.25	1.30	1.35	1.40	1.45	1.50
8.45	1.065	1.053	1.041	1.030	1.019	1.016	1.014	1.016	1.019	1.021
8.50	1.070	1.058	1.046	1.035	1.024	1.021	1.019	1.020	1.023	1.025
8.55	1.075	1.063	1.051	1.040	1.029	1.025	1.023	1.024	1.026	1.028
8.60	1.081	1.068	1.056	1.045	1.034	1.030	1.028	1.028	1.030	1.031
8.65	1.086	1.073	1.061	1.049	1.039	1.034	1.032	1.032	1.034	1.035
8.70	1.092	1.078	1.066	1.054	1.044	1.039	1.037	1.036	1.038	1.039
8.75	1.097	1.083	1.070	1.058	1.049	1.043	1.041	1.040	1.041	1.043
8.80	1.102	1.088	1.075	1.063	1.054	1.048	1.045	1.044	1.045	1.047
8.85	1.108	1.094	1.080	1.068	1.058	1.052	1.049	1.048	1.048	1.049
8.90	1.113	1.099	1.085	1.073	1.063	1.056	1.053	1.052	1.052	1.052
8.95	1.118	1.103	1.090	1.077	1.067	1.060	1.057	1.056	1.056	1.056
9.00	1.124	1.108	1.095	1.082	1.072	1.064	1.061	1.060	1.060	1.060
9.05	1.129	1.113	1.100	1.086	1.076	1.068	1.065	1.064	1.064	1.064
9.10	1.135	1.118	1.105	1.091	1.081	1.073	1.069	1.068	1.068	1.068
9.15	1.140	1.123	1.109	1.096	1.085	1.077	1.073	1.071	1.071	1.071
9.20	1.146	1.128	1.114	1.101	1.090	1.082	1.078	1.075	1.075	1.074
9.25	1.151	1.133	1.119	1.105	1.094	1.086	1.082	1.079	1.078	1.077
9.30	1.157	1.138	1.124	1.110	1.099	1.091	1.086	1.083	1.082	1.081
9.35	1.162	1.143	1.129	1.115	1.104	1.095	1.090	1.087	1.086	1.084
9.40	1.167	1.148	1.134	1.120	1.109	1.100	1.095	1.092	1.090	1.088
9.45	1.172	1.153	1.139	1.124	1.113	1.104	1.099	1.096	1.094	1.092
9.50	1.178	1.159	1.144	1.129	1.118	1.109	1.103	1.100	1.098	1.096
9.55	1.182	1.164	1.149	1.134	1.122	1.113	1.107	1.104	1.101	1.099
9.60	1.189	1.169	1.154	1.139	1.127	1.118	1.111	1.108	1.104	1.102
9.65	1.194	1.174	1.158	1.143	1.132	1.122	1.115	1.112	1.108	1.105

9.70	1.200	1.179	1.163	1.148	1.137	1.127	1.120	1.116	1.112	1.109
9.75	1.205	1.184	1.168	1.153	1.141	1.131	1.124	1.120	1.115	1.113
9.80	1.210	1.189	1.173	1.158	1.146	1.135	1.128	1.124	1.119	1.117
9.85	1.215	1.194	1.178	1.162	1.150	1.139	1.132	1.128	1.123	1.120
9.90	1.221	1.200	1.183	1.167	1.155	1.143	1.137	1.132	1.127	1.123
9.95	1.226	1.205	1.187	1.171	1.159	1.147	1.141	1.136	1.130	1.126
10.00	1.231	1.210	1.192	1.176	1.164	1.152	1.145	1.140	1.134	1.130
10.05	1.236	1.215	1.197	1.180	1.168	1.156	1.149	1.144	1.138	1.134
10.10	1.242	1.220	1.202	1.185	1.173	1.160	1.153	1.148	1.142	1.138
10.15	1.247	1.225	1.206	1.190	1.177	1.164	1.157	1.152	1.145	1.141
10.20	1.252	1.230	1.211	1.195	1.181	1.169	1.162	1.155	1.149	1.145
10.25	1.257	1.235	1.216	1.199	1.185	1.174	1.166	1.159	1.153	1.148
10.30	1.263	1.240	1.221	1.204	1.190	1.179	1.170	1.163	1.157	1.152
10.35	1.268	1.245	1.226	1.208	1.194	1.183	1.174	1.167	1.160	1.156
10.40	1.274	1.250	1.231	1.213	1.199	1.188	1.178	1.171	1.164	1.160
10.45	1.279	1.255	1.236	1.218	1.203	1.192	1.182	1.175	1.168	1.164
10.50	1.285	1.260	1.241	1.223	1.208	1.196	1.186	1.179	1.172	1.168
10.55	1.290	1.265	1.246	1.227	1.212	1.200	1.190	1.183	1.176	1.171
10.60	1.296	1.270	1.251	1.232	1.217	1.204	1.195	1.187	1.180	1.174
10.65	1.301	1.275	1.256	1.236	1.221	1.208	1.199	1.191	1.183	1.177
10.70	1.307	1.280	1.261	1.241	1.226	1.213	1.203	1.195	1.187	1.181
10.75	1.312	1.285	1.265	1.246	1.231	1.217	1.207	1.198	1.191	1.185
10.80	1.318	1.291	1.270	1.251	1.236	1.222	1.211	1.202	1.195	1.189
10.85	1.323	1.296	1.275	1.255	1.240	1.226	1.215	1.206	1.198	1.193
10.90	1.329	1.301	1.280	1.260	1.245	1.230	1.219	1.210	1.202	1.197
10.95	1.334	1.306	1.284	1.265	1.249	1.235	1.223	1.214	1.206	1.200
11.00	1.339	1.311	1.289	1.270	1.254	1.240	1.228	1.218	1.210	1.204
11.05	1.344	1.316	1.294	1.275	1.258	1.244	1.232	1.221	1.213	1.208
11.10	1.350	1.321	1.299	1.280	1.263	1.249	1.236	1.225	1.217	1.211
11.15	1.355	1.326	1.303	1.284	1.268	1.253	1.240	1.229	1.221	1.214

Table A–3, continued

Pseudo-reduced Pressure, p_{pr}	*Pseudoreduced Temperature,* T_{pr}									
	1.60	1.70	1.80	1.90	2.00	2.20	2.40	2.60	2.80	3.00
8.45	1.029	1.041	1.055	1.066	1.076	1.092	1.103	1.112	1.118	1.121
8.50	1.032	1.044	1.058	1.069	1.078	1.094	1.105	1.113	1.119	1.123
8.55	1.035	1.047	1.060	1.071	1.080	1.096	1.107	1.115	1.120	1.124
8.60	1.039	1.050	1.063	1.073	1.082	1.098	1.109	1.117	1.122	1.126
8.65	1.042	1.052	1.065	1.075	1.084	1.099	1.110	1.118	1.123	1.127
8.70	1.046	1.055	1.068	1.078	1.087	1.101	1.112	1.120	1.125	1.129
8.75	1.048	1.058	1.070	1.080	1.089	1.103	1.113	1.121	1.126	1.130
8.80	1.051	1.061	1.073	1.082	1.091	1.105	1.115	1.122	1.127	1.132
8.85	1.055	1.064	1.075	1.084	1.093	1.107	1.116	1.124	1.129	1.133
8.90	1.058	1.067	1.078	1.087	1.095	1.109	1.118	1.126	1.131	1.135
8.95	1.061	1.069	1.080	1.089	1.097	1.111	1.120	1.127	1.132	1.136
9.00	1.064	1.072	1.083	1.092	1.100	1.113	1.122	1.129	1.134	1.138
9.05	1.067	1.075	1.085	1.094	1.102	1.115	1.124	1.130	1.135	1.139
9.10	1.070	1.078	1.088	1.097	1.104	1.117	1.126	1.132	1.137	1.140
9.15	1.074	1.081	1.091	1.099	1.106	1.119	1.128	1.133	1.138	1.141
9.20	1.078	1.084	1.094	1.102	1.108	1.122	1.130	1.135	1.140	1.143
9.25	1.081	1.086	1.096	1.104	1.110	1.124	1.132	1.136	1.141	1.145
9.30	1.084	1.089	1.099	1.106	1.112	1.126	1.134	1.138	1.143	1.147
9.35	1.087	1.092	1.101	1.108	1.114	1.128	1.136	1.140	1.145	1.148
9.40	1.090	1.095	1.104	1.111	1.117	1.130	1.138	1.142	1.147	1.149
9.45	1.093	1.098	1.106	1.113	1.119	1.132	1.140	1.143	1.148	1.150
9.50	1.097	1.102	1.109	1.116	1.122	1.134	1.142	1.145	1.150	1.152
9.55	1.100	1.104	1.111	1.118	1.124	1.136	1.143	1.147	1.151	1.153
9.60	1.103	1.107	1.114	1.121	1.126	1.139	1.145	1.149	1.153	1.155
9.65	1.106	1.110	1.117	1.123	1.128	1.141	1.147	1.150	1.154	1.156

9.70	1.109	1.113	1.120	1.126	1.131	1.143	1.149	1.152	1.156	1.158
9.75	1.113	1.116	1.122	1.128	1.133	1.145	1.150	1.154	1.157	1.159
9.80	1.117	1.119	1.125	1.131	1.136	1.148	1.152	1.156	1.159	1.161
9.85	1.120	1.122	1.127	1.133	1.138	1.150	1.154	1.157	1.160	1.162
9.90	1.123	1.126	1.130	1.135	1.140	1.152	1.156	1.159	1.162	1.164
9.95	1.126	1.129	1.133	1.138	1.142	1.154	1.158	1.161	1.163	1.165
10.00	1.130	1.132	1.136	1.141	1.144	1.156	1.160	1.163	1.165	1.167
10.05	1.133	1.135	1.139	1.143	1.146	1.158	1.162	1.164	1.166	1.168
10.10	1.137	1.138	1.141	1.145	1.149	1.160	1.164	1.166	1.168	1.170
10.15	1.140	1.141	1.144	1.147	1.151	1.162	1.165	1.167	1.169	1.171
10.20	1.143	1.144	1.147	1.150	1.154	1.164	1.167	1.169	1.171	1.173
10.25	1.146	1.147	1.149	1.152	1.156	1.166	1.169	1.171	1.173	1.175
10.30	1.149	1.150	1.152	1.155	1.159	1.168	1.171	1.173	1.175	1.177
10.35	1.152	1.153	1.155	1.157	1.161	1.170	1.173	1.174	1.176	1.178
10.40	1.156	1.156	1.158	1.160	1.164	1.172	1.175	1.176	1.178	1.180
10.45	1.159	1.159	1.160	1.162	1.166	1.174	1.177	1.177	1.179	1.181
10.50	1.162	1.162	1.162	1.165	1.169	1.176	1.179	1.179	1.181	1.183
10.55	1.165	1.165	1.165	1.168	1.171	1.178	1.181	1.181	1.182	1.184
10.60	1.169	1.168	1.168	1.171	1.174	1.181	1.183	1.183	1.184	1.186
10.65	1.172	1.170	1.171	1.174	1.176	1.183	1.184	1.184	1.185	1.187
10.70	1.176	1.173	1.174	1.177	1.179	1.185	1.186	1.186	1.187	1.189
10.75	1.179	1.176	1.176	1.179	1.182	1.187	1.187	1.188	1.189	1.190
10.80	1.182	1.179	1.179	1.182	1.185	1.189	1.189	1.190	1.191	1.192
10.85	1.186	1.182	1.182	1.185	1.187	1.191	1.191	1.192	1.192	1.193
10.90	1.190	1.186	1.186	1.188	1.190	1.193	1.193	1.194	1.194	1.195
10.95	1.193	1.189	1.189	1.191	1.192	1.195	1.195	1.196	1.196	1.197
11.00	1.196	1.193	1.192	1.194	1.195	1.197	1.197	1.198	1.198	1.199
11.05	1.199	1.196	1.194	1.196	1.197	1.199	1.199	1.199	1.199	1.200
11.10	1.202	1.199	1.196	1.199	1.200	1.201	1.201	1.201	1.201	1.202
11.15	1.206	1.202	1.199	1.201	1.202	1.203	1.203	1.203	1.203	1.203

Table A–3, continued

Pseudo-reduced Pressure, p_{pr}	*Pseudoreduced Temperature,* T_{pr}									
	1.05	1.10	1.15	1.20	1.25	1.30	1.35	1.40	1.45	1.50
11.20	1.360	1.331	1.308	1.289	1.273	1.257	1.245	1.233	1.225	1.218
11.25	1.365	1.336	1.313	1.294	1.277	1.261	1.249	1.237	1.228	1.221
11.30	1.370	1.341	1.318	1.299	1.282	1.265	1.253	1.241	1.232	1.225
11.35	1.375	1.346	1.323	1.303	1.286	1.269	1.257	1.245	1.236	1.228
11.40	1.381	1.351	1.328	1.308	1.291	1.273	1.261	1.249	1.240	1.232
11.45	1.386	1.356	1.333	1.313	1.295	1.277	1.265	1.253	1.243	1.236
11.50	1.391	1.362	1.338	1.318	1.300	1.282	1.269	1.257	1.247	1.240
11.55	1.396	1.367	1.343	1.322	1.304	1.286	1.273	1.260	1.251	1.244
11.60	1.402	1.372	1.348	1.327	1.309	1.291	1.277	1.264	1.255	1.248
11.65	1.407	1.377	1.353	1.331	1.313	1.295	1.281	1.268	1.258	1.251
11.70	1.412	1.382	1.358	1.336	1.317	1.300	1.286	1.272	1.262	1.255
11.75	1.417	1.387	1.363	1.340	1.321	1.305	1.288	1.276	1.265	1.259
11.80	1.423	1.392	1.368	1.345	1.326	1.309	1.294	1.280	1.269	1.262
11.85	1.428	1.397	1.373	1.350	1.330	1.313	1.298	1.284	1.273	1.266
11.90	1.433	1.402	1.378	1.355	1.335	1.318	1.302	1.288	1.277	1.270
11.95	1.438	1.407	1.383	1.359	1.339	1.322	1.306	1.292	1.280	1.274
12.00	1.444	1.413	1.388	1.364	1.344	1.327	1.310	1.296	1.284	1.278
12.05	1.449	1.418	1.393	1.368	1.348	1.331	1.314	1.299	1.288	1.281
12.10	1.454	1.423	1.398	1.373	1.353	1.335	1.318	1.303	1.292	1.285
12.15	1.459	1.428	1.403	1.378	1.357	1.339	1.322	1.307	1.296	1.288
12.20	1.465	1.433	1.408	1.383	1.362	1.344	1.326	1.311	1.300	1.292
12.25	1.470	1.438	1.413	1.388	1.366	1.348	1.330	1.315	1.304	1.295
12.30	1.475	1.443	1.418	1.392	1.371	1.353	1.334	1.319	1.307	1.299
12.35	1.480	1.448	1.423	1.397	1.375	1.357	1.338	1.323	1.311	1.303
12.40	1.486	1.453	1.428	1.402	1.380	1.362	1.342	1.327	1.315	1.307

12.45	1.491	1.458	1.433	1.407	1.384	1.366	1.346	1.330	1.319	1.310
12.50	1.496	1.464	1.438	1.411	1.389	1.370	1.350	1.334	1.323	1.313
12.55	1.501	1.469	1.443	1.416	1.394	1.374	1.354	1.338	1.326	1.317
12.60	1.507	1.474	1.448	1.420	1.399	1.379	1.358	1.342	1.330	1.321
12.65	1.512	1.479	1.453	1.425	1.404	1.383	1.362	1.346	1.334	1.325
12.70	1.517	1.484	1.458	1.430	1.409	1.388	1.367	1.350	1.337	1.329
12.75	1.522	1.489	1.463	1.435	1.413	1.392	1.371	1.354	1.341	1.332
12.80	1.527	1.495	1.468	1.439	1.418	1.397	1.375	1.358	1.345	1.336
12.85	1.532	1.500	1.473	1.444	1.422	1.401	1.379	1.362	1.348	1.339
12.90	1.537	1.505	1.477	1.449	1.427	1.405	1.383	1.365	1.352	1.342
12.95	1.542	1.510	1.482	1.454	1.431	1.409	1.387	1.369	1.356	1.346
13.00	1.548	1.515	1.487	1.458	1.436	1.414	1.391	1.374	1.360	1.350
13.05	1.553	1.520	1.492	1.463	1.440	1.418	1.395	1.378	1.364	1.354
13.10	1.558	1.525	1.497	1.468	1.445	1.422	1.399	1.381	1.367	1.358
13.15	1.563	1.530	1.502	1.473	1.449	1.426	1.403	1.385	1.371	1.361
13.20	1.568	1.535	1.507	1.477	1.453	1.431	1.408	1.389	1.374	1.364
13.25	1.573	1.540	1.512	1.482	1.458	1.435	1.412	1.393	1.378	1.368
13.30	1.578	1.546	1.517	1.486	1.463	1.440	1.416	1.398	1.382	1.372
13.35	1.583	1.551	1.521	1.491	1.467	1.444	1.420	1.401	1.385	1.375
13.40	1.589	1.556	1.526	1.495	1.472	1.449	1.424	1.404	1.389	1.379
13.45	1.594	1.561	1.531	1.500	1.476	1.453	1.428	1.408	1.393	1.382
13.50	1.599	1.566	1.536	1.504	1.481	1.457	1.432	1.412	1.397	1.386
13.55	1.604	1.571	1.541	1.509	1.485	1.461	1.436	1.416	1.400	1.389
13.60	1.609	1.577	1.546	1.514	1.490	1.466	1.440	1.420	1.404	1.393
13.65	1.614	1.582	1.551	1.518	1.494	1.470	1.444	1.424	1.408	1.396
13.70	1.619	1.587	1.556	1.523	1.499	1.475	1.448	1.428	1.412	1.400
13.75	1.624	1.592	1.561	1.527	1.503	1.479	1.452	1.432	1.416	1.404
13.80	1.630	1.597	1.566	1.532	1.508	1.483	1.456	1.436	1.419	1.408
13.85	1.635	1.602	1.571	1.537	1.513	1.487	1.460	1.439	1.423	1.411
13.90	1.640	1.608	1.575	1.542	1.517	1.492	1.465	1.443	1.426	1.414

Table A–3, continued

Pseudo-reduced Pressure, p_{pr}	*Pseudoreduced Temperature,* T_{pr}									
	1.60	1.70	1.80	1.90	2.00	2.20	2.40	2.60	2.80	3.00
11.20	1.210	1.205	1.202	1.204	1.204	1.205	1.205	1.205	1.205	1.205
11.25	1.213	1.208	1.205	1.206	1.206	1.207	1.207	1.207	1.207	1.207
11.30	1.217	1.211	1.208	1.209	1.209	1.209	1.209	1.209	1.209	1.209
11.35	1.220	1.214	1.211	1.211	1.211	1.211	1.211	1.211	1.210	1.210
11.40	1.224	1.217	1.214	1.214	1.214	1.214	1.214	1.213	1.212	1.212
11.45	1.227	1.220	1.217	1.216	1.216	1.216	1.216	1.215	1.214	1.213
11.50	1.231	1.224	1.220	1.219	1.219	1.219	1.218	1.217	1.216	1.215
11.55	1.234	1.227	1.223	1.222	1.221	1.221	1.220	1.219	1.218	1.217
11.60	1.238	1.230	1.226	1.225	1.224	1.223	1.222	1.221	1.220	1.219
11.65	1.241	1.233	1.229	1.227	1.226	1.225	1.223	1.222	1.221	1.220
11.70	1.245	1.236	1.232	1.230	1.229	1.227	1.225	1.224	1.223	1.222
11.75	1.248	1.239	1.235	1.233	1.232	1.229	1.227	1.226	1.225	1.224
11.80	1.252	1.243	1.238	1.236	1.234	1.231	1.229	1.228	1.227	1.226
11.85	1.255	1.246	1.240	1.238	1.236	1.233	1.231	1.229	1.228	1.227
11.90	1.259	1.250	1.243	1.241	1.239	1.236	1.233	1.231	1.230	1.229
11.95	1.262	1.253	1.246	1.244	1.241	1.238	1.235	1.233	1.231	1.230
12.00	1.266	1.256	1.249	1.247	1.244	1.240	1.237	1.235	1.233	1.232
12.05	1.269	1.259	1.252	1.249	1.246	1.242	1.238	1.236	1.235	1.234
12.10	1.272	1.262	1.255	1.251	1.249	1.244	1.240	1.238	1.237	1.236
12.15	1.275	1.265	1.258	1.254	1.251	1.246	1.242	1.240	1.238	1.237
12.20	1.279	1.268	1.261	1.257	1.254	1.248	1.244	1.242	1.240	1.239
12.25	1.283	1.271	1.264	1.259	1.256	1.250	1.246	1.244	1.242	1.241
12.30	1.287	1.274	1.267	1.262	1.259	1.252	1.248	1.246	1.244	1.243
12.35	1.290	1.277	1.270	1.265	1.261	1.254	1.250	1.247	1.246	1.245
12.40	1.293	1.281	1.273	1.268	1.264	1.256	1.252	1.249	1.248	1.247

12.45	1.296	1.284	1.276	1.270	1.266	1.258	1.254	1.250	1.249	1.248
12.50	1.300	1.288	1.279	1.273	1.269	1.261	1.256	1.252	1.251	1.250
12.55	1.303	1.290	1.282	1.275	1.271	1.263	1.258	1.254	1.252	1.251
12.60	1.307	1.293	1.285	1.278	1.273	1.265	1.260	1.256	1.254	1.253
12.65	1.310	1.297	1.288	1.280	1.276	1.267	1.261	1.257	1.256	1.255
12.70	1.314	1.300	1.291	1.283	1.279	1.269	1.263	1.259	1.258	1.257
12.75	1.317	1.303	1.294	1.286	1.281	1.271	1.265	1.261	1.259	1.258
12.80	1.321	1.306	1.297	1.289	1.283	1.273	1.267	1.263	1.261	1.260
12.85	1.324	1.309	1.299	1.291	1.286	1.275	1.269	1.265	1.263	1.261
12.90	1.328	1.312	1.302	1.294	1.289	1.277	1.271	1.267	1.265	1.263
12.95	1.331	1.315	1.305	1.297	1.291	1.279	1.272	1.268	1.266	1.265
13.00	1.335	1.319	1.308	1.300	1.293	1.281	1.274	1.270	1.268	1.267
13.05	1.338	1.321	1.310	1.302	1.295	1.283	1.276	1.271	1.269	1.268
13.10	1.341	1.324	1.313	1.304	1.298	1.286	1.278	1.273	1.271	1.270
13.15	1.345	1.327	1.316	1.307	1.300	1.288	1.280	1.275	1.273	1.271
13.20	1.349	1.331	1.320	1.310	1.303	1.290	1.282	1.277	1.275	1.273
13.25	1.352	1.334	1.322	1.313	1.305	1.292	1.284	1.278	1.276	1.275
13.30	1.355	1.337	1.325	1.316	1.308	1.294	1.286	1.280	1.278	1.277
13.35	1.358	1.340	1.328	1.318	1.310	1.296	1.287	1.281	1.279	1.278
13.40	1.362	1.343	1.331	1.321	1.313	1.298	1.289	1.283	1.281	1.280
13.45	1.365	1.346	1.334	1.323	1.315	1.300	1.291	1.285	1.283	1.281
13.50	1.369	1.349	1.337	1.326	1.317	1.302	1.293	1.287	1.285	1.283
13.55	1.372	1.352	1.340	1.328	1.319	1.304	1.295	1.289	1.287	1.285
13.60	1.376	1.355	1.343	1.331	1.322	1.306	1.297	1.291	1.289	1.287
13.65	1.379	1.358	1.346	1.334	1.324	1.308	1.298	1.292	1.290	1.288
13.70	1.382	1.361	1.349	1.337	1.327	1.310	1.300	1.294	1.292	1.290
13.75	1.385	1.364	1.351	1.339	1.329	1.312	1.301	1.296	1.293	1.291
13.80	1.389	1.368	1.354	1.342	1.332	1.314	1.303	1.298	1.295	1.293
13.85	1.393	1.371	1.357	1.345	1.334	1.316	1.305	1.300	1.297	1.295
13.90	1.397	1.374	1.360	1.348	1.337	1.318	1.307	1.301	1.299	1.297

Table A–3, continued

Pseudo-reduced Pressure, p_{pr}	*Pseudoreduced Temperature,* T_{pr}									
	1.05	1.10	1.15	1.20	1.25	1.30	1.35	1.40	1.45	1.50
13.95	1.645	1.613	1.580	1.546	1.522	1.496	1.469	1.447	1.430	1.417
14.00	1.650	1.618	1.585	1.550	1.526	1.500	1.473	1.451	1.434	1.421
14.05	1.655	1.623	1.590	1.555	1.531	1.504	1.478	1.455	1.437	1.425
14.10	1.661	1.628	1.595	1.560	1.535	1.509	1.482	1.459	1.441	1.429
14.15	1.666	1.633	1.600	1.565	1.539	1.513	1.486	1.463	1.445	1.432
14.20	1.671	1.639	1.605	1.570	1.544	1.518	1.490	1.467	1.449	1.436
14.25	1.676	1.644	1.609	1.575	1.548	1.522	1.494	1.471	1.452	1.439
14.30	1.681	1.649	1.614	1.580	1.553	1.527	1.498	1.475	1.456	1.443
14.35	1.686	1.650	1.619	1.584	1.557	1.531	1.502	1.478	1.460	1.446
14.40	1.692	1.659	1.624	1.589	1.561	1.536	1.506	1.482	1.464	1.450
14.45	1.697	1.664	1.629	1.594	1.565	1.540	1.510	1.486	1.468	1.454
14.50	1.702	1.669	1.634	1.598	1.570	1.544	1.515	1.490	1.472	1.458
14.55	1.707	1.674	1.639	1.603	1.575	1.548	1.520	1.494	1.476	1.461
14.60	1.712	1.679	1.643	1.608	1.580	1.552	1.524	1.498	1.480	1.465
14.65	1.717	1.684	1.648	1.613	1.584	1.556	1.528	1.502	1.484	1.468
14.70	1.722	1.690	1.653	1.617	1.589	1.561	1.532	1.505	1.488	1.472
14.75	1.727	1.695	1.658	1.622	1.594	1.565	1.536	1.509	1.491	1.475
14.80	1.733	1.700	1.663	1.627	1.598	1.570	1.540	1.513	1.495	1.479
14.85	1.738	1.705	1.668	1.632	1.603	1.574	1.544	1.517	1.498	1.483
14.90	1.743	1.710	1.673	1.636	1.607	1.579	1.549	1.521	1.502	1.487
14.95	1.748	1.715	1.678	1.641	1.612	1.583	1.553	1.525	1.505	1.490
15.00	1.753	1.720	1.682	1.645	1.616	1.588	1.558	1.529	1.508	1.493

	1.60	1.70	1.80	1.90	2.00	2.20	2.40	2.60	2.80	3.00
13.95	1.399	1.377	1.363	1.350	1.339	1.320	1.309	1.302	1.300	1.298
14.00	1.402	1.380	1.366	1.353	1.341	1.322	1.311	1.304	1.302	1.300
14.05	1.405	1.383	1.369	1.356	1.343	1.324	1.313	1.306	1.303	1.301
14.10	1.409	1.386	1.372	1.358	1.346	1.326	1.315	1.308	1.305	1.303
14.15	1.412	1.389	1.375	1.360	1.348	1.328	1.317	1.309	1.307	1.305
14.20	1.416	1.392	1.378	1.363	1.351	1.330	1.319	1.311	1.309	1.307
14.25	1.419	1.395	1.381	1.366	1.353	1.332	1.320	1.313	1.310	1.308
14.30	1.423	1.398	1.384	1.369	1.356	1.334	1.322	1.315	1.312	1.310
14.35	1.426	1.401	1.387	1.371	1.358	1.336	1.323	1.316	1.313	1.311
14.40	1.430	1.404	1.390	1.374	1.360	1.338	1.325	1.318	1.315	1.313
14.45	1.433	1.407	1.393	1.376	1.362	1.340	1.327	1.320	1.317	1.315
14.50	1.437	1.411	1.396	1.379	1.365	1.342	1.329	1.322	1.319	1.317
14.55	1.440	1.414	1.398	1.382	1.367	1.344	1.331	1.324	1.321	1.318
14.60	1.443	1.417	1.400	1.385	1.370	1.346	1.333	1.326	1.323	1.320
14.65	1.446	1.420	1.403	1.387	1.372	1.348	1.334	1.327	1.324	1.321
14.70	1.450	1.423	1.407	1.390	1.375	1.350	1.336	1.329	1.326	1.323
14.75	1.453	1.426	1.410	1.392	1.377	1.352	1.338	1.330	1.327	1.325
14.80	1.457	1.429	1.413	1.395	1.380	1.354	1.340	1.332	1.329	1.327
14.85	1.460	1.432	1.416	1.398	1.382	1.356	1.341	1.334	1.331	1.328
14.90	1.463	1.436	1.419	1.401	1.385	1.358	1.343	1.336	1.333	1.330
14.95	1.466	1.439	1.421	1.403	1.387	1.360	1.345	1.338	1.334	1.331
15.00	1.470	1.442	1.424	1.406	1.390	1.362	1.347	1.340	1.336	1.333

APPENDIX B.
Gas Measurement Factors

TABLE B–1.

Basic Orifice Factors, F_b, for Flange Taps, Mcfd/$\sqrt{h_w p_m}$ *

p_b = *14.65 psia* T_m = *520°R or 60°F*
T_b = *520°R or 60°F* γ_g = *1.000*
All other factors = *1.000*

Nominal Diameter (in.)	*2*			*3*				*4*	
ASA Schedule	*160*	*80*	*40*	*xxHvy*	*160*	*80*	*40*	*xxHvy*	*160*
Inside Diameter (in.)	*1.689*	*1.939*	*2.067* †	*2.300*	*2.626*	*2.900*	*3.068* †	*3.152*	*3.438*
Orifice Diameter (in.)									
0.250	0.3064	0.3066	0.3067	0.3068	0.3068	0.3067	0.3066	0.3065	0.3064
0.375	0.6871	0.6863	0.6860	0.6856	0.6852	0.6849	0.6848	0.6847	0.6845
0.500	1.225	1.221	1.219	1.217	1.215	1.214	1.214	1.213	1.213
0.625	1.933	1.919	1.914	1.908	1.902	1.899	1.897	1.897	1.895
0.750	2.826	2.790	2.779	2.764	2.751	2.744	2.740	2.739	2.735
0.875	3.932	3.851	3.824	3.792	3.765	3.750	3.744	3.741	3.733
1.000	5.303	5.127	5.073	5.006	4.951	4.924	4.912	4.907	4.893
1.125	7.022	6.665	6.557	6.428	6.324	6.237	6.251	6.242	6.217
1.250	9.310	8.533	8.329	8.087	7.901	7.810	7.771	7.755	7.713
1.375	—	10.82	10.46	10.03	9.705	9.551	9.486	9.459	9.388
1.500	—	—	13.09	12.33	11.78	11.52	11.41	11.37	11.25
1.625	—	—	—	15.06	14.16	13.75	13.58	13.51	13.33
1.750	—	—	—	—	16.92	16.28	16.01	15.90	15.63
1.875	—	—	—	—	20.15	19.16	18.75	18.59	18.18
2.000	—	—	—	—	—	22.46	21.86	21.62	21.01

* Adapted by the Interstate Oil Compact Commission from *American Gas Association Gas Measurement Committee Report No. 3.* Used by permission of the Interstate Oil Compact Commission.

† Use if only the nominal diameter is known.

Table B–1, continued

Nominal Diameter (in.)	*4*		*6*				*8*		
ASA Schedule	*80*	*40*	*xxHvy*	*160*	*80*	*40*	*80*	*40*	*30*
Inside Diameter (in.)	*3.826*	*4.026*	*4.897*	*5.189*	*5.761*	*6.065*	*7.625*	*7.981*	*8.071*
Orifice Diameter (in.)		†				†			
2.125	—	—	—	—	—	26.33	25.40	25.05	24.17
2.250	—	—	—	—	—	—	29.52	28.96	27.70
2.375	—	—	—	—	—	—	—	—	31.65
2.500	—	—	—	—	—	—	—	—	36.16
0.250	0.3062	0.3061							
0.375	0.6842	0.6841							
0.500	1.212	1.212	1.211	1.211	1.211	1.211			
0.625	1.893	1.892	1.890	1.890	1.889	1.889			
0.750	2.731	2.729	2.724	2.723	2.721	2.720			
0.875	3.726	3.723	3.713	3.711	3.707	3.706	3.700	3.700	3.700
1.000	4.879	4.874	4.859	4.855	4.850	4.847	4.838	4.836	4.836
1.125	6.194	6.186	6.161	6.156	6.147	6.143	6.129	6.127	6.126
1.250	7.675	7.661	7.622	7.613	7.600	7.595	7.575	7.572	7.571
1.375	9.326	9.303	9.242	9.230	9.211	9.203	9.176	9.172	9.171
1.500	11.16	11.12	11.03	11.01	10.98	10.97	10.93	10.93	10.93
1.625	13.17	13.12	12.98	12.95	12.91	12.90	12.84	12.84	12.84
1.750	15.39	15.31	15.10	15.06	15.01	14.98	14.91	14.90	14.90
1.875	17.83	17.71	17.40	17.34	17.26	17.23	17.14	17.13	17.13
2.000	20.50	20.32	19.88	19.80	19.69	19.65	19.53	19.51	19.51
2.125	23.43	23.18	22.56	22.45	22.30	22.24	22.07	22.05	22.05
2.250	26.66	26.30	25.44	25.29	25.09	25.[illegible]1	24.79	24.76	24.75
2.375	30.22	29.72	28.55	28.34	28.06	27.[illegible]5	27.66	27.62	27.61
2.500	34.15	33.48	31.88	31.60	31.22	31.09	30.70	30.65	30.64
2.625	38.50	37.60	35.45	35.09	34.59	34.41	33.92	33.85	33.84

2.750	43.37	42.15	39.31	38.82	38.18	37.94	37.30	37.22	37.21
2.875	—	47.19	43.46	42.82	41.99	41.69	40.87	40.77	40.74
3.000	—	52.97	47.94	47.12	46.04	45.66	44.61	44.49	44.46
3.125	—	—	52.78	51.72	50.35	49.86	48.54	48.39	48.36
3.250	—	—	58.02	56.68	54.94	54.32	52.67	52.48	52.44
3.375	—	—	63.70	62.01	59.83	59.05	56.99	56.76	56.71
3.500	—	—	69.87	67.77	65.04	64.04	64.07	61.24	61.17
3.625	—	—	76.76	73.97	70.60	69.40	66.28	65.92	65.85
3.750	—	—	—	80.73	76.55	75.08	71.25	70.82	70.73
3.875	—	—	—	88.27	82.91	81.12	76.46	75.94	75.83
4.000	—	—	—	—	89.73	87.56	81.91	81.29	81.15
4.250	—	—	—	—	105.01	101.8	93.62	92.72	92.53
4.500	—	—	—	—	—	118.3	106.5	105.2	105.0
4.750	—	—	—	—	—	—	120.7	118.9	118.5
5.000	—	—	—	—	—	—	136.3	134.0	133.4
5.250	—	—	—	—	—	—	153.7	150.5	149.8
5.500	—	—	—	—	—	—	173.0	168.7	167.8
5.750	—	—	—	—	—	—	—	189.0	187.7
6.000	—	—	—	—	—	—	—	—	210.1

† Use if only the nominal diameter is known.

TABLE B–2.

Basic Orifice Factors, F_b, for Pipe Taps, Mcfd/$\sqrt{h_w p_m}$ *

p_b = *14.65 psia* T_m = *520°R or 60°F*
T_b = *520°R or 60°F* γ_g = *1.000*
All other factors = 1.000

Nominal Diameter (in.)	*2*			*3*				*4*	
ASA Schedule	*160*	*80*	*40*	*xxHvy*	*160*	*80*	*40*	*xxHvy*	*160*
Inside Diameter (in.)	*1.689*	*1.939*	*2.067*	*2.300*	*2.626*	*2.900*	*3.068*	*3.152*	*3.438*
Orifice Diameter (in.)			†				†		
0.250	0.3101	0.3092	0.3089	0.3085	0.3080	0.3078	0.3076	0.3076	0.3074
0.375	0.7085	0.7022	0.6999	0.6970	0.6943	0.6928	0.6921	0.6918	0.6910
0.500	1.296	1.275	1.266	1.255	1.245	1.239	1.237	1.235	1.232
0.625	2.105	2.049	2.029	2.001	2.974	1.958	1.951	1.948	1.939
0.750	3.191	3.061	3.016	2.955	2.897	2.864	2.848	2.840	2.820
0.875	4.651	4.368	4.273	4.149	4.036	3.972	3.941	3.928	3.889
1.000	6.647	6.060	5.871	5.630	5.419	5.303	5.249	5.226	5.159
1.125	9.458	8.277	7.915	7.467	7.090	6.889	6.797	6.756	6.645
1.250	—	11.24	10.57	9.762	9.106	8.770	8.618	8.554	8.375
1.375	—	—	14.09	12.66	11.55	11.00	10.76	10.65	10.37
1.500	—	—	—	16.39	14.54	13.65	13.27	13.11	12.68
1.625	—	—	—	—	18.23	16.83	16.24	15.99	15.34
1.750	—	—	—	—	22.85	20.67	19.77	19.40	18.43
1.875	—	—	—	—	—	25.35	23.99	23.44	22.01
2.000	—	—	—	—	—	31.15	29.09	28.28	26.20
2.125	—	—	—	—	—	—	35.36	34.15	31.12
2.250	—	—	—	—	—	—	—	—	36.97
2.375	—	—	—	—	—	—	—	—	43.99

Table B–2, continued

Nominal Diameter (in.)	*4*		*6*				*8*		
ASA Schedule	*80*	*40*	*xxHvy*	*160*	*80*	*40*	*80*	*40*	*30*
Inside Diameter (in.)	*3.826*	*4.026*	*4.897*	*5.189*	*5.761*	*6.065*	*7.625*	*7.981*	*8.071*
Orifice Diameter (in.)		†				†			
0.250	0.3071	0.3070							
0.375	0.6901	0.6898							
0.500	1.229	1.228	1.224	1.224	1.222	1.222			
0.625	1.930	1.927	1.917	1.915	1.912	1.910			
0.750	2.801	2.793	2.771	2.766	2.759	2.756			
0.875	3.851	3.836	3.791	3.782	3.768	3.762	3.743	3.740	3.739
1.000	5.093	5.066	4.986	4.969	4.943	4.933	4.899	4.894	4.893
1.125	6.537	6.494	6.364	6.335	6.291	6.274	6.217	6.209	6.207
1.250	8.202	8.134	7.933	7.888	7.819	7.791	7.701	7.687	7.685
1.375	10.11	10.00	9.703	9.636	9.534	9.492	9.354	9.334	9.330
1.500	12.28	12.12	11.68	11.59	11.44	11.38	11.18	11.15	11.15
1.625	14.75	14.52	13.89	13.76	13.56	13.47	13.19	13.15	13.14
1.750	17.56	17.23	16.35	16.16	15.88	15.77	15.38	15.33	15.32
1.875	20.76	20.30	19.06	18.81	18.43	18.28	17.77	17.69	17.68
2.000	24.41	23.77	22.07	21.73	21.22	21.02	20.35	20.25	20.23
2.125	28.60	27.71	25.39	24.93	24.27	24.00	23.14	23.01	22.98
2.250	33.43	32.20	29.07	28.46	27.59	27.24	26.13	25.98	25.94
2.375	39.02	37.34	33.14	32.34	31.21	30.76	29.35	29.15	29.11
2.500	45.55	43.25	37.66	36.61	35.15	34.57	32.80	32.55	32.49
2.625	53.24	50.10	42.67	41.32	39.44	38.71	36.49	36.18	36.11

* Adapted by the Interstate Oil Compact Commission from *American Gas Association Gas Measurement Committee Report No. 3.* Used by permission of the Interstate Oil Compact Commission.

† Use if only the nominal diameter is known.

Table B–2, continued

Nominal Diameter (in.)	*4*		*6*				*8*		
ASA Schedule	*80*	*40*	*xxHvy*	*160*	*80*	*40*	*80*	*40*	*30*
Inside Diameter (in.)	*3.826*	*4.026*	*4.897*	*5.189*	*5.761*	*6.065*	*7.625*	*7.981*	*8.071*
Orifice Diameter (in.)		†				†			
2.750	—	58.09	48.26	46.52	44.12	43.20	40.43	40.05	39.97
2.875	—	—	54.50	52.27	49.23	48.08	44.64	44.18	44.07
3.000	—	—	61.50	58.65	54.81	53.37	49.13	48.57	48.44
3.125	—	—	69.38	65.75	60.91	59.13	53.93	53.24	53.09
3.250	—	—	78.30	73.67	67.61	65.40	59.04	58.22	58.03
3.375	—	—	88.46	82.55	74.98	72.23	64.49	63.51	63.28
3.500	—	—	—	92.56	83.09	79.71	70.31	69.13	68.87
3.625	—	—	—	103.9	92.05	87.90	76.52	75.12	74.80
3.750	—	—	—	—	102.0	96.89	83.16	81.48	81.11
3.875	—	—	—	—	113.1	106.8	90.25	88.27	87.82
4.000	—	—	—	—	125.4	117.7	97.84	95.49	94.96
4.250	—	—	—	—	—	—	114.7	111.4	110.7
4.500	—	—	—	—	—	—	134.0	129.6	128.6
4.750	—	—	—	—	—	—	156.5	150.4	149.0
5.000	—	—	—	—	—	—	182.7	174.3	172.5
5.250	—	—	—	—	—	—	213.6	202.1	199.7
5.500	—	—	—	—	—	—	—	234.7	231.3

† Use if only the nominal diameter is known.

TABLE B–3.

Basic Critical Flow Prover Factors, F_p, Mcfd/psia *

$$p_b = 14.65 \text{ psia}$$
$$T_b = 520°R \text{ or } 60°F$$
$$T_m = 520°R \text{ or } 60°F$$
$$\gamma_g = 1.000$$

2-in. Prover		*4-in. Prover*	
Orifice Diameter (in.)	*Factor, F_p*	*Orifice Diameter (in.)*	*Factor, F_p*
1⁄16	0.06569	1⁄4	1.074
3⁄32	0.1446	3⁄8	2.414
1⁄8	0.2716	1⁄2	4.319
3⁄16	0.6237	5⁄8	6.729
7⁄32	0.8608	3⁄4	9.643
1⁄4	1.115	7⁄8	13.11
5⁄16	1.714	1	17.08
3⁄8	2.439	1 1⁄8	21.52
7⁄16	3.495	1 1⁄4	26.57
1⁄2	4.388	1 3⁄8	31.99
5⁄8	6.638	1 1⁄2	38.12
3⁄4	9.694	1 3⁄4	52.07
7⁄8	13.33	2	68.80
1	17.53	2 1⁄4	88.19
1 1⁄8	22.45	2 1⁄2	110.6
1 1⁄4	28.34	2 3⁄4	136.9
1 3⁄8	34.82	3	168.3
1 1⁄2	43.19		

* Adapted from *Bureau of Mines Monograph 7* by the Interstate Oil Compact Commission. Used by permission of the Interstate Oil Compact Commission.

TABLE B–4.

*Flowing Temperature Factors, F_t**

$$F_t = \sqrt{\frac{520}{T_m}}$$

Observed Tempera-ture, °F	*0*	*1*	*2*	*3*	*4*	*5*	*6*	*7*	*8*	*9*
0	1.063	1.062	1.061	1.060	1.059	1.057	1.056	1.055	1.054	1.053
10	1.052	1.051	1.050	1.049	1.047	1.046	1.045	1.044	1.043	1.042
20	1.041	1.040	1.039	1.038	1.037	1.035	1.034	1.033	1.032	1.031
30	1.030	1.029	1.028	1.027	1.026	1.025	1.024	1.023	1.022	1.021
40	1.020	1.019	1.018	1.017	1.016	1.015	1.014	1.013	1.012	1.011
50	1.010	1.009	1.008	1.007	1.006	1.005	1.004	1.003	1.002	1.001
60	1.000	0.9990	0.9981	0.9971	0.9962	0.9952	0.9943	0.9933	0.9924	0.9915
70	0.9905	0.9896	0.9887	0.9877	0.9868	0.9859	0.9850	0.9840	0.9831	0.9822
80	0.9813	0.9804	0.9795	0.9786	0.9777	0.9768	0.8759	0.9750	0.9741	0.9732
90	0.9723	0.9715	0.9706	0.9697	0.9688	0.9680	0.9671	0.9662	0.9653	0.9645

100	0.9636	0.9628	0.9619	0.9610	0.9602	0.9594	0.9585	0.9577	0.9568	0.9560
110	0.9551	0.9543	0.9535	0.9526	0.6518	0.9510	0.9501	0.9493	0.9485	0.9477
120	0.9469	0.9460	0.9452	0.9444	0.9436	0.9428	0.9420	0.9412	0.9404	0.9396
130	0.9388	0.9380	0.9372	0.9364	0.9356	0.9349	0.9341	0.9333	0.9325	0.9317
140	0.9309	0.9302	0.9294	0.9286	0.9279	0.9271	0.9263	0.9256	0.9248	0.9240
150	0.9233	0.9225	0.9217	0.9210	0.9202	0.9195	0.9187	0.9180	0.9173	0.9165
160	0.9158	0.9150	0.9143	0.9135	0.9128	0.9121	0.9114	0.9106	0.9099	0.9092
170	0.9085	0.9077	0.9071	0.9063	0.9055	0.9048	0.9042	0.9035	0.9028	0.9020
180	0.9014	0.9007	0.9000	0.8992	0.8985	0.8979	0.8972	0.8965	0.8958	0.8951
190	0.8944	0.8937	0.8931	0.8923	0.8916	0.8910	0.8903	0.8896	0.8889	0.8882
200	0.8876	0.8870	0.8863	0.8856	0.8849	0.8843	0.8836	0.8830	0.8823	0.8816
210	0.8810	0.8803	0.8797	0.8790	0.8784	0.8777	0.8770	0.8764	0.8758	0.8751
220	0.8745	0.8738	0.8732	0.8725	0.8719	0.8713	0.8706	0.8700	0.8694	0.8687
230	0.8681	0.8675	0.8668	0.8662	0.8656	0.8650	0.8644	0.8637	0.8631	0.8625
240	0.8619	0.8613	0.8606	0.8600	0.8594	0.8588	0.8582	0.8576	0.8570	0.8564

* Used by permission of the Interstate Oil Compact Commission.

TABLE B–5.

*Specific Gravity Factors, F_g**

$$F_g = \sqrt{\frac{1}{\gamma_g}}$$

Specific Gravity γ_g	0.000	0.001	0.002	0.003	0.004	0.005	0.006	0.007	0.008	0.009
0.550	1.348	1.347	1.346	1.345	1.344	1.342	1.341	1.340	1.339	1.338
0.560	1.336	1.335	1.334	1.333	1.332	1.330	1.329	1.328	1.327	1.326
0.570	1.325	1.323	1.322	1.321	1.320	1.319	1.318	1.316	1.315	1.314
0.580	1.313	1.312	1.311	1.310	1.309	1.307	1.306	1.305	1.304	1.303
0.590	1.302	1.301	1.300	1.299	1.298	1.296	1.295	1.294	1.293	1.292
0.600	1.291	1.290	1.289	1.288	1.287	1.286	1.285	1.284	1.282	1.281
0.610	1.280	1.279	1.278	1.277	1.276	1.275	1.274	1.273	1.272	1.271
0.620	1.270	1.269	1.268	1.267	1.266	1.265	1.264	1.263	1.262	1.261
0.630	1.260	1.259	1.258	1.257	1.256	1.255	1.254	1.253	1.252	1.251
0.640	1.250	1.249	1.248	1.247	1.246	1.245	1.244	1.243	1.242	1.241
0.650	1.240	1.239	1.238	1.237	1.237	1.236	1.235	1.234	1.233	1.232
0.660	1.231	1.230	1.229	1.228	1.227	1.226	1.225	1.224	1.224	1.223
0.670	1.222	1.221	1.220	1.219	1.218	1.217	1.216	1.215	1.214	1.214
0.680	1.213	1.212	1.211	1.210	1.209	1.208	1.207	1.206	1.206	1.205
0.690	1.204	1.203	1.202	1.201	1.200	1.200	1.199	1.198	1.197	1.196
0.700	1.195	1.194	1.194	1.193	1.192	1.191	1.190	1.189	1.188	1.188
0.710	1.187	1.186	1.185	1.184	1.183	1.183	1.182	1.181	1.180	1.179
0.720	1.179	1.178	1.177	1.176	1.175	1.174	1.174	1.173	1.172	1.171
0.730	1.170	1.170	1.169	1.168	1.167	1.166	1.166	1.165	1.164	1.163
0.740	1.162	1.162	1.161	1.160	1.159	1.159	1.158	1.157	1.156	1.155

0.750	1.155	1.154	1.153	1.152	1.152	1.151	1.150	1.149	1.149	1.148
0.760	1.147	1.146	1.146	1.145	1.144	1.143	1.143	1.142	1.141	1.140
0.770	1.140	1.139	1.138	1.137	1.137	1.136	1.135	1.134	1.134	1.133
0.780	1.132	1.132	1.131	1.130	1.129	1.129	1.128	1.127	1.127	1.126
0.790	1.125	1.124	1.124	1.123	1.122	1.122	1.121	1.120	1.119	1.119
0.800	1.118	1.117	1.117	1.116	1.115	1.115	1.114	1.113	1.112	1.112
0.810	1.111	1.110	1.110	1.109	1.108	1.108	1.107	1.106	1.106	1.105
0.820	1.104	1.104	1.103	1.102	1.102	1.101	1.100	1.100	1.099	1.098
0.830	1.098	1.097	1.096	1.096	1.095	1.094	1.094	1.093	1.092	1.092
0.840	1.091	1.090	1.090	1.089	1.089	1.088	1.087	1.087	1.086	1.085
0.850	1.085	1.084	1.083	1.083	1.082	1.081	1.081	1.080	1.080	1.079
0.860	1.078	1.078	1.077	1.076	1.076	1.075	1.075	1.074	1.073	1.073
0.870	1.072	1.072	1.071	1.070	1.070	1.069	1.068	1.068	1.067	1.067
0.880	1.066	1.065	1.065	1.064	1.064	1.063	1.062	1.062	1.061	1.061
0.890	1.606	1.059	1.059	1.058	1.058	1.057	1.056	1.056	1.055	1.055
0.900	1.054	1.054	1.053	1.052	1.052	1.051	1.051	1.050	1.049	1.049
0.910	1.048	1.048	1.047	1.047	1.046	1.045	1.045	1.044	1.044	1.043
0.920	1.043	1.042	1.041	1.041	1.040	1.040	1.039	1.039	1.038	1.038
0.930	1.037	1.036	1.036	1.035	1.035	1.034	1.034	1.033	1.033	1.032
0.940	1.031	1.031	1.030	1.030	1.029	1.029	1.028	1.028	1.027	1.027
0.950	1.026	1.025	1.025	1.024	1.024	1.023	1.023	1.022	1.022	1.021
0.960	1.021	1.020	1.020	1.019	1.019	1.018	1.017	1.017	1.016	1.016
0.970	1.015	1.015	1.014	1.014	1.013	1.013	1.012	1.012	1.011	1.011
0.980	1.010	1.010	1.009	1.009	1.008	1.008	1.007	1.007	1.006	1.006
0.990	1.005	1.005	1.004	1.004	1.003	1.003	1.002	1.002	1.001	1.001

* Used by permission of the Interstate Oil Compact Commission.

INDEX